Maria T. Bailey

Mom 3.0

Marketing *with* Today's Mothers by Leveraging New Media & Technology

Wyatt-MacKenzie Publishing, Inc.
DEADWOOD, OREGON

Mom 3.0

Marketing *with* Today's Mothers by Leveraging New Media & Technology

FIRST EDITION

ISBN: 978-1-932279-90-0

Library of Congress Control Number: 2008929178

www.WyMacPublishing.com (541) 964-3314

Requests for permission or further information should be addressed to:
Wyatt-MacKenzie Publishing, 15115 Highway 36, Deadwood, Oregon 97430

Printed in United States of America.

Dedication

To my four wonderful children Madison, Owen, Keenan and Morgan and my husband, Tim.

Thank you for allowing me to create a career around motherhood.

Table of Contents

I would like to thank Michael Mendenhall at HP for sharing his personal insights and thought leadership on marketing and brand reputational management in a digital economy, based on his 20+ years of industry experience.

Michael Mendenhall is senior vice president and chief marketing officer at HP, where he directs all aspects of the company's corporate marketing operations globally. The organization oversees brand strategy, internal and external communications, digital engagement, global citizenship, integrated design, customer intelligence, regional marketing and hp.com. He is a member of the World Economic Forum's Global Agenda Council on Marketing and Branding, the Academy of Television Arts & Sciences, the senior advisory board of the Executive Marketing Council, and the Marketing 50, a club of 50 top non-competitive chief marketing officer peers who come together for strategic collaboration.

Foreword

"YOU EITHER ARE ONE, HAVE ONE, OR KNOW ONE!" WAS ONE OF THE VERY first statements that Maria Bailey mentioned to me when I was first introduced to her several years ago. Not surprisingly, Maria was talking about Moms and that single expression was one of the most compelling, simplistic "elevator pitches" I have heard in my entire career. I mean let's be honest, how could I possibly argue with that?

The power and elegance of moms has always been, is currently, and will continue to be a beautiful thing. For too long, moms have been lumped under this vast, blanket persona of being just another subsection of consumers. Maria powerfully distilled this notion in her groundbreaking book, *Trillion Dollar Moms*, and has somehow found a way to exceed expectations once again and highlight one of the most fascinating examples of convergence we are seeing in the marketplace today. A convergence of mothers and technology that has unlimited implications for all consumers, businesses, marketers, mothers, fathers, sons, daughters, and oh yeah, the Internet as we know it today.

We are in the middle of a technological revolution that has resulted in a fundamental paradigm shift in the overall media landscape which has had profound implications on how companies today manage their relationships with key stakeholders, manage their reputation, and manage their brand. The era of brands being built solely with a traditional 15, 30, and/or 60 second TV spot is dead. Brands today are now being built in the digital net space (think mobile, Video on Demand, web, social networks, online communities, etc.) as opposed to just traditional channels such as TV, radio, and print.

Consumer behavior is in constant change and very dynamic. However, the biggest change today is that technology has increased the pace of that change exponentially leaving many companies, agencies, and strategies fundamentally ill equipped and out of date. I can't tell you how many times I have been involved in meetings with marketing and advertising agencies that think first and foremost about television, and commence their agendas accordingly.

Moms are paving the way to what will soon be an entirely new experience of what we traditionally know today as the Internet. We are moving towards an Internet experience, in fact a whole digital experience that will one day be personalized top to bottom…and I don't mean customized, I mean PERSONALized. The entire digital space will be proactive in nature, as opposed to reactive as it is today, based off what site we choose to visit, what term we choose to search for, or what community we choose to join. We will be encompassed in a digital conversation, updated in real time that will be two-way as oppose to one-way. We have just started to scratch the surface of what this next generation of digital experiences will look like and Maria has magnificently captured the actions of mothers today that are prime examples of the foundation this digital 3.0 experience will embody.

And I don't want to underestimate the impact that Mothers around the world (this is a true global movement) are having. Mothers are not only participating in this digital revolution, they are helping to define it. Moms are not just engaged in these emerging digital mediums, they are helping to direct their future. Maria was at the forefront of disrupting the marketplace and challenging industry convention with *Trillion Dollar Moms* and forcing organizations and marketers to rethink how they interact and view moms as a consumer base. *Mom 3.0* is another glaring wake-up call of the influence and change that moms all around the world are driving in this global digital economy.

I strongly believe that all companies, regardless of size, geography, industry, or heritage, must think FIRST about how to develop a comprehensive digital strategy and how their brand, their communications, and their marketing will play out in this digital space. The lines of internal communications and external communications are blurring. Companies must recognize that the walled garden of in-house messages has crumbled and what is said internally must be assumed will be disseminated externally.

While your doors may close at 5:00, your phones may turn off at 8:00, and your TV ads may stop running at 9:00, your online digital presence doesn't operate on a 40-hour work week. It's always ON which reinforces the power and simultaneous opportunity of providing platforms, solutions, forums, and new ways for individuals to interact and engage with your brand and your products and services when they want, how they want, and where they want.

Research is showing that the typical 24 hour day around the world has now eclipsed the 40 hour mark and in some cases is now close to 50 hours all due in part to multi-tasking. There is no better example of the power of multi-tasking than looking at the lifestyles and challenges that mothers today are facing every day. She may not look at a single TV screen all day or pick up a single newspaper or magazine. The 30 second TV spot that you spent weeks crafting through all stages of strategy, ideation, execution and launch has gone completely unnoticed. The only medium she may be interacting with is the Internet, and now, your website which has constantly been pushed to the side (or even worse, neglected all together!) is the sole interaction she is having with your brand, the sole influencing aspect of a purchase decision, and the sole personification of your company's reputation. If that doesn't make you think twice, I don't know what will.

Organizations constantly focus on the "structured" influencers of Media, Press, and Analysts (and don't get me wrong, they are and always will be immensely important). But organizations often miss a fundamental aspect of building a successful reputational management practice in a digital economy which must also incorporate the "unstructured" influencers. Moms are one of the most influential and prominent examples of this unstructured group that has historically been overlooked.

Positive word of mouth marketing has always been the pinnacle of a successful marketing campaign or marketing strategy, and so too has negative word of mouth marketing been the dreaded signal you may want to dust off that resume. Moms have always been the most trusted source of information and recommendations in the neighborhood or community. And now coupled with technology, the ability of mothers to help other mothers with decisions and ideas and the ability of mothers to communicate with other mothers around the world has never been greater.

We also know mothers across the board are the ones typically responsible for managing and controlling the finances in a household. But now, millions upon millions of moms have an endless array of technology at their fingertips to directly influence brands, influence organizations' reputations, and directly influence *thousands* of other moms' purchasing decisions at the click of a mouse. This is playing out every day, every minute, and every second in this digital net space and Maria will walk you through how, why, and where this revolution is taking place.

Maria is a successful CEO, a dear friend and most importantly a terrific mother and it is with great humility and excitement that I say this book will force all organizations, all marketing and advertising agencies, and all employees to rethink how they have traditionally operated and structured their organizations as we witness the power of Mom 3.0.

Michael Mendenhall

Introduction

THE BOOK YOU'RE HOLDING IS ONE THAT MY FORMER PUBLISHER didn't want to publish. Wait, let me explain. They told me that the "Marketing to Moms" market was already saturated with books, written by none other than yours truly. Talk about shooting yourself in the foot! What they didn't understand was that since I wrote my first book *Marketing to Moms: Getting Your Share of the Trillion Dollar Market* (Random House, 2002), moms have gone from launching websites to launching social networks, and now instead of posting responses on a message board, they are posting on blogs.

My former publishing company might have been right that I've saturated the market with books about my passion; however, they were wrong about the need for a new book. A great deal has changed in the strategies and tactics in connecting with moms even since they published *Trillion Dollar Moms: Marketing to a New Generation of Mothers* (Dearborn, 2005). There are only a few of the same programs today that I recommend to clients that we executed for clients five years ago.

As I watched the behaviors of mothers change over the past few years, new media emerged with technological advances. Marketing professionals inquired about how to leverage new media and technology together to tap the Mom Market. The publishers may not be ready for it, but I know from the phone calls and emails I receive every day that marketers are eager for more. My intent with *Mom 3.0: Marketing with Mothers using New Media and Technology* (Wyatt-MacKenzie, 2008) is to provide marketers, not only with the how-to's of effectively tapping the Mom Market, but also give a glimpse into what's ahead in marketing with moms.

Why *Mom 3.0*? Because I believe mothers are always ahead of society and, most of the time, ahead of where marketers believe they are. Think about it; moms have been a powerful group since the 1950's yet it wasn't until they elected Bill Clinton to office that the media took notice of their political power. It wasn't until I wrote *Marketing to Moms* in 2001 that anyone, including our government, took the time to calculate their spending power. Today that very number my team calculated, $1.7 trillion, is so widely used that it rarely gets attributed to me. Fortunately, for the bottom line of many companies, millions are now being spent more wisely and more targeted to tapping a mom's consumer power.

I will demonstrate in this book that moms are exhibiting behaviors that catapult the current state of technology. They are also marketing your products and brands in ways that few companies are doing themselves. My intention in writing *Mom 3.0* was to give today's mother a more defining identity so that marketers can understand how she is bridging together technology and everyday behaviors to create channels of communication, solutions and ecosystems that can be integrated with brand messages.

Meet Mom 3.0. She is a powerful consumer who not only purchases products, but influences the decision-making process of her peers through the use of new media and content that is relevant and intuitive, and delivers her, and her peers, ecosystems of solutions. I know there are a lot of words in my definition, however, by focusing on a few key words you will come to fully understand the contents of this book.

First, we will briefly examine mom's spending power and consumer behaviors. What is she buying and how is she purchasing it? Today she's doing more than pushing a cart down the aisle as social selling, online purchases and peer recommendations grow in popularity, and as Blackberry-toting twittering Millennial women join the ranks of motherhood. The purchases she is making are often a part of an ecosystem she's created and may include multiple products, brands or services bundled together to create a solution in her life. You will learn how to become a part of her ecosystems.

As technology and functionality change and mothers are leveraging the time-saving abilities, marketers need to continue to be a part of the equation and keep up with the pace of on-the-go moms. In each chapter I will provide strategies and tactics that companies can leverage to stay connected with mom's consumer patterns. Today's mothers are more frequently involving their peers in their purchasing behaviors.

We will learn how to identify the Mom Mavens or Influencers who can carry your brand message into online communities, social networks, blogs and beyond. Examples of companies leveraging these influencers will provide insights into programs marketers can execute to successfully sell product or services to mothers near and far. I've taken each new media and technology and included never-before-released research on its function within the Mom Market.

Additionally, we will examine how mothers are currently using the technology or media in everyday life, how marketers are leveraging it to deliver marketing messages, and the opportunities for you to update your marketing initiatives to maximize both.

Throughout the book we will address the final issue which is content. Whether you are communicating with moms via Vcast, mobile devices, podcast or blog, the relevance of your message is paramount. With each tactic, I will give examples of applying the technology to content that is valuable, timely and, most importantly, develops a dialogue with mothers.

If there is one takeaway that I hope to give the readers of this book, it too is found in the title. The key word here is "with." Unlike the titles and subtitle of my two previous books, I am encouraging marketers to market WITH moms rather than TO moms. The marketing pendulum has shifted and more and more moms feel empowered to control the marketing and communication that companies produce. Mom bloggers are conducting product reviews which are followed by thousands of readers, and Twitterites with toddlers are sending tweets with great deals-of-the-week attached.

Inside social networks, moms are comparing experiences and referring peers to companies that provide superb customer service. And when businesses fall short on brand promises, these moms are blogging and vlogging about their less-than-positive experience with your product. Just as obtaining a mom's seal of approval can send your sales up, one negative blog post or tweet to followers can send your sales plummeting. In the world of consumer generated media and a generation of mothers who are adopting new technologies at a record rate, it's imperative to engage with these women to spread your marketing messages.

Mom 3.0 is filled with ideas, insights and creative programs to help you update your current strategies and prepare your brand for the future of mom marketing.

Moms are engaged in using new technology, which means it must be a significant part of your marketing plan if you want to play in her sandbox.

CHAPTER

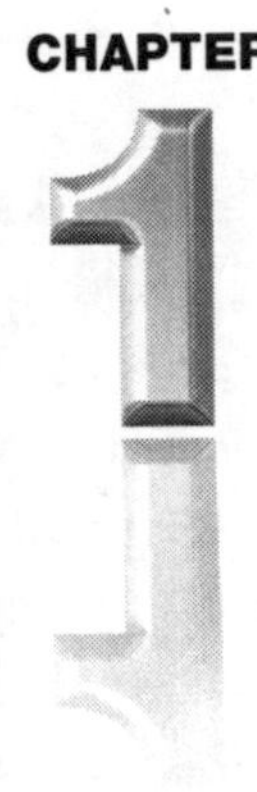

Meet Mom 3.0

MEET JENN. SHE IS A MOTHER, WIFE, BUSINESS OWNER, AND AUTHOR. She uses digital photography to archive family events, market her business and share ideas with other momprenuers. In her purse, she carries a Blackberry, cell phone and iPod loaded with the latest Mom Talk Radio show. Her family is quite active and when it comes to keeping their schedules in sync, Jenn relies on Yahoo! Calendar. Dinner is always a challenge so when she's at a loss for new recipes, her e-friends supply dinner ideas in virtual social groups on MomJunction or Clubmom. Jenn is a member of several networking groups, all with women she's never met face-to-face; however many she calls her best friends. She can rattle off a list of other moms who don't share her zip code but share her parenting style, her business acumen, or her favorite past time. She describes her life as integrated—changing between her multiple roles minute by minute.

Her minivan is much more than a means of transportation. It's an environment that supports this chameleon-like lifestyle. Armed with her wireless card and laptop, the front seat serves as her office while waiting in the parking lot for the afternoon bell to ring at which time the back seat is transformed into a mobile homework station while driving to the multitudes of after-school activities. In between soccer practice and an early evening baseball game, the minivan becomes her family's dining room compliments of Outback Curbside Service or Subway, depending on the day of the week.

Only a mega multi-tasker with the most extraordinary logistical systems in place could manage the demands of Jenn's life. In fact, many have written books to try to answer the question that seems to define today's modern mother, "How does she do it all?" The answer lies in two simple words and the conversion of two important tools—technology and systems. Ironically they are the two key elements of the emergence of Web 3.0. In fact, as the development of Web 3.0 perplexes developers, researchers, and the leading web designers, today's mothers are living the experience of Web 3.0 every day. Meet Jenn. She is Mom 3.0.

What exactly is Mom 3.0?

Mom 3.0 is the living, breathing, personification of what computer geeks foresee in the emergence of Web 3.0. In fact, today's mother could be called the blueprint of tomorrow's Internet. To best understand this statement, it is necessary to understand the expectations or predictions of what the next generation of Internet will offer and how individuals will interact with it. According to a recent *New York Times* article, start-up companies along with leaders such as Google and Yahoo! are clamoring to develop a new layer on today's Web. This new dimension will ultimately make the Internet less of a catalogue of information and more of a guide by adding some level of human reasoning. Instead of a user searching for a vacation destination appropriate for a family with toddlers and teens by clicking through hundreds of options, Web 3.0 would apply deductions to the available content on the Internet to present only the most relevant vacation opportunities. Web 3.0 will be able to evaluate user-generated comments within blogs, message boards, and podcasts, and calculate the value of the content to the users' needs. It may even be able to match the personality of a blogger to the personality of a reader, offering a virtual introduction between the two. Doesn't this sound like someone you know? Doesn't it describe a person who can process lots of information, apply emotion to product recommendations to create reaction by other consumers, and deliver content among peers in a timely, valuable, and relevant manner? It describes today's mom.

It's often been hypothesized that mothers, or women in general, have a different brain configuration than men explaining their innate ability to multi-task. If one believes this theory it may validate that a secondary layer of the Internet is absolutely necessary to create Web 3.0. Today's mother is a walking database of information similar to today's Internet. Her mind is filled with factoids from Grandma's wedding cookie recipe to the easiest cure for colic to the fastest way to please a man when she is too tired to say "no." Like Web 2.0, she shares this cataloged information on a daily basis via written word, audio clips, videos, and just chatting around the playground. Offering a second layer of information delivery will turn Web 2.0 into a multi-tasker in the form of Web 3.0.

The Mom Market

Fortunately for marketers, the emergence of Web 3.0 will collide naturally with the most lucrative consumer group on Earth—the Mom Market. Women with children control $2 trillion in spending each year in the United States alone. Think about that—$2 trillion. That's more than the gross national product of Australia, the Philippines, or Portugal. The purchases that these women are making are as diverse as the ways in which they are managing their families. Mothers control 85% of the household income and buy everything from new cars to life insurance to diapers to men's deodorant. They are Chief Memory Officers, Chief Care Givers, and Chief Household Managers, among other roles. However, in a survey conducted by BSM Media of 600 mothers across the US, 72% said the title they would most commonly give themselves was Chief Financial Officer.

The economic power of these women can also be felt well beyond their living rooms and kitchens. Forty-two percent of boomer moms are involved in some type of elder care. Fifty-seven percent said they frequently make purchases for extended family members. This means that they are not only controlling 85% of their household's income, they are also controlling a percent of other household incomes. In addition to household income, today's mothers are running businesses and community organizations.

To fully understand the impacts of this financial power, take a look at the number of female owned businesses in the US. According to Patricia Cobe, who co-authored the book *Mompreneurs Online* and trademarked the term "mompreneur," an estimated 10 million mothers are running home-based businesses while juggling the demands of children. Many of these women hold on to their identity as stay-at-home moms so they go uncalculated by the government as female business owners. The impact of moms in the business world is one of the greatest untapped opportunities for companies who may still view moms as single-dimension consumers. Finally, it's important not to forget the 26 million traditionally working mothers who leave the home each day to pursue careers in an office away from their family. Many of these women act as purchasing agents for their companies as well as for their homes.

Why is it important to recognize the multiple consumer channels of the Mom Market? Aside from the obvious impact it can make to your bottom line, it's important because moms want you to recognize them. In a survey by the Marketing to Moms Coalition, when asked how marketers should connect with them, moms overwhelmingly said, "I want marketers to recognize my multiple roles and the many hats I wear." It speaks to a woman's desire to feel appreciated and respected. Moms want to know that you realize the challenges they have with their children, the intelligence they possess in business, the sexiness they strive for in spite of the lack of time to exercise, and the struggle they have in communicating with their husbands. Mothers nurture relationships and they want relationships not only with their children, family and friends, but with the companies and brands who earn their business.

The good news for you as a marketer is that if you successfully establish a relationship, mothers are very loyal consumers. In fact, according to BSM Media research, 90% of moms will stay with the same brand if the product meets their expectations. Even more important to their multi-dimensional lives, 92% will purchase the same products and brands for home and office. This means multiple channels of consumption for your brand and your products.

Ironically marketing specifically to moms is a relatively new target market for most companies. It wasn't until the release of my first book *Marketing to Moms: Getting Your Share of the Trillion Dollar Market* (Dearborn, 2002) that anyone, including the US government, had quantified the annual spending of mothers. Sure, toys, diapers and some household cleaning products companies wrote commercials and print ads directed at moms but very few banks, car manufacturers or even food companies targeted mothers. Thankfully, companies such as Best Buy, Hewlett-Packard, Whirlpool and Precious Moments have stepped forward to devote resources to tapping the Mom Market. Their growth and increased sales have brought a heightened awareness to this important consumer segment. Many of the successful programs implemented by these Fortune 100 brands have been incorporated in this book.

New Technology

Moms are speaking to each other with many of today's new technological advances. They are producing podcasts with shows that focus on particular lifestyles such as Mojo Moms, who are self-proclaimed seekers of intelligent content, and Creative Moms, who are interested in non-traditional parenting choices. Mothers are chronicling their journey through motherhood in blogs and sharing their children's daily events with video. A large number of moms are taking know-how and knowledge, and mixing it with technology, to become self-made, self-published experts. Combine a book with a blog, website, virtual audience and podcast and even the most neophyte marketer would agree that you have a powerful marketing machine. This is what moms are doing every day at record numbers.

Recently BSM Media conducted an extensive study of moms and new technology. The online survey polled 600 mothers of varying demographic and psychographic profiles. While only 29% of moms have actually blogged, over 60% have read a blog either occasionally or regularly. Thirty-five percent have listened to a podcast with 23% of these women tuning in to one on a regular basis. Most of the activity between new technology and moms seems to be in video content. Over 79% of mothers say they have watched a video online. Interestingly, they allow their children to watch them as well, without any guidelines currently in place to manage the length of time the children view the videos. This is important to marketers because if moms have not established rules in regards to videos it means that she is open to their consumption in her home. It is a unique opportunity to connect with her while she is defining her emotions toward this medium.

When asked what kinds of videos they prefer, 49% of moms said content that includes solutions or advice from experts. Fourteen percent say they enjoy videos by other moms and 20% want the video to be professionally produced. An overwhelming 94% of moms have posted content online with replies to message boards being the most common source of activity; blogs and articles followed close behind. Photos play an active role in their engagement online. Fifty-three percent say they have uploaded digital images online. Finally, when asked what kind of new technology interests them most, blogs barely beat out videos by one percent.

It's clear that moms are engaged in using new technology, which means it *must* be a significant part of your marketing plan if you want to play in her sandbox. There are several companies who are doing a good job at adapting their marketing messages to moms in these new ways. Perhaps some of the work already being done will spark ideas for embracing new technologies as part of your new marketing strategy.

Companies Connecting with Moms through New Marketing

Whirlpool

Whirlpool has taken an innovate approach to connecting with moms through new marketing techniques. Ultimately, Whirlpool wants to sell washers and dryers; however they utilize new marketing strategies to connect with moms. I invite you to visit www.whirlpoolappliances.ca or www.whirlpool.com to see firsthand the multidimensional marketing platform that they are delivering in the Mom Market. You might be surprised to find more than the expected online catalog and store locator. Instead you will discover a robust site that connects with potential customers by establishing a relevant conversation that supports the individual lifestyles of today's mothers.

Among the product information, there is a section called "Find Balance" and here is where the conversation begins. Whirlpool links itself to motherhood by recognizing that mothers seek balance. The company knows that one way in which a mom might find her sense of balance is by getting the laundry done. So it's a connection that makes perfect sense in the mind of their audience. Once on the Find Balance channel, mothers can find tips on "saying no," "managing life changes" and "how to find time in the day." On Whirlpool's US site they take the theme of balance even further. Here moms can listen to their weekly podcast, The American Family. The podcast is hosted by Audrey Reed-Granger, director of marketing and public relations.

As Whirlpool describes it, The American Family is a discussion-based podcast that addresses matters that impact families with diverse backgrounds and experiences, featuring real, everyday people and subject-matter experts. Utilizing a podcast works on many levels for Whirlpool. First, it provides Whirlpool with a tool for an ongoing conversation with moms that is valuable and relevant. The content provides mothers with the much sought after knowledge and know-how that positions them with their peers. It's a vehicle of communication that can travel with the mother at her pace, whether she's jogging behind a stroller or riding in her minivan. She doesn't need an appointment to speak with Whirlpool but instead can manage the time and relationship based on her schedule. Whirlpool's message is available to her when she wants it and where she wants it, on her terms. It says to a busy mom, "We understand your lifestyle and want to be a part of it." It's also a tool that she can share with other moms by passing along her iPod downloads or sending a link to girlfriends. The use of a podcast also creates an informal community of moms who share a common interest in the topic of the day. Although these moms don't technically interact, there is the perception that others are listening to the show.

There are other advantages and benefits to Whirlpool beyond direct communication with customers. Producing a podcast allows them to distribute their marketing messages beyond their website. You can find The American Family podcast on Podcastalley.com, iTunes and many of the other common podcast directories. In searches engines, it puts the Whirlpool brand on pages beyond appliance searches.

Whirlpool continues their dialogue with moms through another innovative marketing initiative. It's called the Whirlpool Brand Mother of Invention Grant contest. It is an annual contest that recognizes mom-owned companies and awards grants to help them grow. The copy describing the contest reads, "Whirlpool Brand knows moms like you are identifying challenges and coming up with innovative and inventive solutions every day. To recognize this motherly ingenuity, the 2006 Whirlpool Brand Mother of Invention Grant is giving moms a chance to turn their business dreams into reality." For the over 5 million momprenuers in the US today, it says, "Whirlpool understands moms, their dreams and their need for help." Additionally it says that Whirlpool respects a mom's ability to do great things. The contest engages moms in an activity that empowers her and creates a meaningful two-way dialogue. Furthermore, it creates a community where moms share business ideas with each other.

An appliance site…what an unlikely place to find a community of moms talking about their companies. However if you speak to any momprenuer, she is likely to tell you that she knows of Whirlpool's Mother of Invention Grant. More important than the traffic that the contest creates for Whirlpool's website is the viral marketing it produces for the appliance company. As we've already discussed, moms love to share and the Mother of Invention Grant gives moms something to share with other moms. It's not uncommon as the deadline approaches to see emails circulating among online communities encouraging moms to compete in the competition. It resonates with moms because less than 2% of all venture capital funding is awarded to female owned businesses. Ironically most moms fund their ideas with personal savings even though the segment launches more businesses than any other group today. To have a company such as Whirlpool recognize their challenges and provide a possible solution goes a long way to solidifying moms' preference for appliances.

If connecting with Whirlpool through audio or community activities isn't enough for moms, there's more. This time the brand delivers a unique experience to potential customers through two marketing channels; video and physical interaction. Whirlpool's Insperience Studio is a hands-on experience that allows women to cook, touch and play with appliances in a non-retail environment. The studio is located in Atlanta, Georgia. However if the Peach state isn't on your list of destinations, Whirlpool allows women to visit via online video. It's a full sensory experience that connects with moms through multiple media channels with messages that form a strong emotional connection. It's new marketing at its best.

Build-A-Bear Workshop

There are other companies who are utilizing new marketing techniques to connect with mothers. Build-A-Bear Workshop successfully connects with moms by delivering customizable products and personalized interactions with the brand. It begins with the ability to customize their main product—a teddy bear. The value of customization certainly can be seen in the rapid growth of this company. Who would have thought a decade ago that a multi-million dollar company could be built by selling teddy bears? Yet, the ability to create not only a customized bear but also a unique retail experience has made this one of the most well known brands among moms today. The success of Build-A-Bear Workshop's marketing is the delivery of consistent messaging through multiple channels, although somewhat different than Whirlpool.

Build-A-Bear connects with moms through a rich loyalty program that rewards mothers and their children for purchases with a robust online shopping experience that presents products based on lifestyle and gifting needs. Their licensing programs have taken the brand into areas that make sense to their core consumer. For instance in early 2007, Build-A-Bear Workshop partnered with Pulaski Furniture to create the Build-A-Bear Home Collection, a full line of children's bedroom furniture that, like its bears, can be customized to the style of the mom and child. Through this licensing agreement, moms can now interact with the brand in a new space and outside of the retail environment. To make the relationship even more valuable to moms, Pulaski and Build-A-Bear gathered a team of mom and child advisors who helped design the furniture. In the end, both companies were not only assured that the product was right for the market but were able to leverage the mom interaction in a very successful public relations effort.

Newbaby.com

A final example of new marketing at work is Newbaby.com, a newly launched website that is best described as "YouTube meets MySpace for mothers." Newbaby is literally Mom 3.0 at its best. Mothers are able to share, learn and connect via video, podcasts, blogs and virtual playgroups. Furthermore, the source of these mediums is provided by both subject experts as well as peers. The experience is definitely multi-sensory, which is a perfect match for the information-hungry audience Newbaby.com draws in. Expectant and new moms consume more information than women at any other life stage. For marketers who want to attract this very lucrative sub-segment of mothers, it's worth delivering your messages through multiple channels in different forms.

Expectant and new moms on average will spend over $10,000 in the first year of their child's life. Latino moms exceed that amount. In an audio and video environment, marketers have the ability to integrate product information into trailers that precede videos or position relevant products within content—taking product placement to a new level. Ultimately there is the ability to link to online purchasing within the content delivery experience, taking the consumer from the learning mode to the purchasing stage. The entire process makes it easy for moms to learn and shop at the same time.

New Media and Moms

One could easily conclude that a new market segment like Mom 3.0 demands new marketing techniques. However, mothers have been engaged in the techniques of new media as fast as they can be created. It wasn't the fate of timing that brought these two together, rather the demands of innate behaviors of mothers. Put simply, new marketing techniques such as blogs, vlogs and social networking meet the communication needs and behavioral habits of the Mom Market. In fact if you were to put a physical face on these new marketing initiatives, the outcome would look like today's mothers.

Think for a minute about what we know about the emotional motivators of blogs and social networks—nurturing relationships, maintaining meaningful dialogues, relevant engagement, and moving in a lot of directions and expending a great deal of energy 24/7. Doesn't this describe the daily role of the mothers you know? She's always on the move, managing to be where she needs to be and when, delivering solutions and recreating herself depending on her audience and nurturing meaningful relationships along the way. Today's mothers are required to balance numerous relationships, all with different personalities and desires, and to customize each of their experiences with her.

Let's put a face and personality to www.YouTube.com to illustrate the point. This media-rich destination plays a lot of different roles to a lot of different people, just as a mother does each and every day. To some YouTube might be a source of entertainment like a bored toddler whining for his mother's attention. To others YouTube may be a source of marketing, selling products, or building a name for the next Tila Tequila. Is this any different from today's pageant mother or sports enthusiast who pushes her child to be the star among their peers? Finally, YouTube can be the source of answers for everyday problems provided by the latest self-made expert. It's really no different than the Mom Maven who shares her parenting expertise with moms sitting around a sandbox at the park.

You see, new media and moms share the same personalities and traits making them the perfect partners in communicating and managing their busy lives. Each serves a large audience while still personalizing the experience for each of the members within the audience. In her role as a mother she nurtures children, each of whom have their own characteristics, individuality and desire to express their own personalities. As a home-based business owner, she may need to warp into the CEO while simultaneously operating as the mail room clerk. In the community, this same mother might take on the role of carnival manager, sports announcer or taxi cab driver shuffling kids from one location to another. In each of these instances throughout her day, she is nurturing relationships while fulfilling a need in her customer whether it's her children, her spouse, clients or community. She is the living, breathing, walking and talking form of new media.

For marketers the challenge is finding that perfect fit with your target mom so that it is inherent for her to carry on a relationship with you, whether that means buying your product or dialing your number when she needs your service. Moms want you to customize your relationship with her while maintaining what feels like a meaningful one-to-one relationship.

You may be asking yourself, "How can these women be considered 3.0 when Web 3.0 doesn't even exist yet?" Good question. The answer can be found in the special talent most moms possess—ecosystem building. It's similar to a beaver building a dam. The animal uses many different types of limbs, branches and vegetation to create a structure that allows him to manage his environment. Moms are no different. If the bridge isn't built or the dam doesn't stop the water, they find a way to create a solution by building an ecosystem of solutions. When it comes to technology, today's mothers are pasting together Web 2.0 functionality with her offline behaviors to create Web 3.0 tools and solutions.

Ecosystems

Moms not only personify Web 3.0 through their behaviors, they walk the walk of integrated communication, information and marketing every single day. Today's mothers are busy. If you ask most moms they will tell you that they are always on the go. In fact, more than 80% of mothers will tell you they need a 25th hour in their day to complete their work. They are on the go non-stop and firing on all cylinders to find solutions while nurturing the people around them. In their quest to achieve their goals they use many different tactics, just like a marketing professional effectively utilizing integrated new media. We know that wireless devices, home PCs, iPods, both terrestrial and satellite radio, and the Internet are just some of the tools in her toolbox. We call these mothers, Digimoms™ because they are leveraging technology to create personal solutions to the everyday challenges and chaos. It's not uncommon today to see a mom managing team websites that allow her to share event photos, solicit snacks for upcoming games or manage the transportation needs of her children.

These technology-savvy moms are also using digital photography to create unique memory books, party invitations and personalized gifts. In their cars while patiently waiting in carpool lines, they are listening to podcasts downloaded to their iPods and syncing up their Blackberries with other moms on the sidelines. Perhaps one of the most important concepts for marketers to understand about these behaviors is that moms are creating ecosystems of solutions. They are pasting together technology and products to create procedures and processes that allow them to manage their busy lives.

As a company who wants moms to hear your message, use your products and help spread your word, you have two choices. You can either be a part of her ecosystem by integrating your brand into her solution-driven world or you can sit on the sideline and hope she selects your product to be a part of her plan. There is no doubt that the former is far more lucrative for your business today and far more likely to establish a tighter bond with the consumer. It's the way to secure tomorrow's sales.

The creation and use of ecosystems is such a powerful tool in the lives of mothers that I devote an entire chapter to this behavior. I have named the tactic of meeting a mom's desire for ecosystems of solutions, Ecosystem Selling™. You will see examples of companies effectively utilizing Ecosystem Selling by bundling products, unifying marketing messages and customizing the customer experience. These modifications to your company's marketing approach can make the relationship with today's moms far more relevant and valuable. Ecosystems allow your products to keep up with today's fast-paced mother, no matter where she is during her day. Additionally as moms move, your marketing messages must move with her. For this reason, success in marketing with moms must be as integrated as she is in her transition between roles.

New Technology and Moms

Your marketing plan must evolve as Internet technology evolves in order to keep up with Mom 3.0. If you have relied on print ads and a website for the last decade, it's time to evolve or get left behind. Moms are integrating new technologies to communicate and nurture relationships. Mothers have been a driving force behind the growth of the Internet. In fact, according to Pew research, women made up 52% of Internet users in 2006. What's changed however since the Internet became mainstream is how moms are using their access to the World Wide Web. In the late nineties, the Internet became a source of quick information. If her child was wetting the bed, a mom could go online and find an article written by an expert on solving this problem. If the family was taking a vacation, she could go online to purchase tickets, reserve hotel rooms and map out their adventure. The Internet was a handy dandy go-to guide for many, if not all, of her know-how needs. It was also a way for her to stay connected with family and friends. She utilized emails, posted messages on boards and shared ideas with moms in chat rooms.

As the comfort level of technology increased for Boomer moms and Gen X and Millennial women began to have children, moms expanded their use of technology to include not only gathering information, but disseminating it as well. She began organizing and distributing photos, managing family schedules with online calendars and shopping for family necessities. For a brief period of time she was even doing her grocery shopping online. The Internet also helped mom find balance between work and family demands. Moms have redefined the traditional lifestyle segmentation of "stay at home" and "working" with new categories with classifications such as "hybrid," "home based working" and "momprenuer." All of these are the outcome of moms utilizing the Internet to create solutions for their lifestyle.

In *Trillion Dollar Moms: Marketing to a New Generation of Mothers* (Dearborn, 2002), we examined the generational make-up of today's Mom Market. Historically it is the first time that three distinct generations of mothers co-exist. Each of these cohorts has their own defining qualities, parenting styles and consumer behaviors. We will review these distinguishing factors in the next chapter as well as build out specific sub-segments of mothers, both for the benefit of those who have not read *Trillion Dollar Moms* and because the mom market is so rapidly changing.

Today, although it has largely gone unrecognized, moms are fueling Web 3.0. In their constant quest for solutions and relationships as well as their propensity to create ecosystems, moms are turning the Web into a tool of empowerment. They are using websites to showcase their children's activities, e-newsletters to build their businesses, blogs to educate others on products and services and podcasts to spread their point of view. Why would mothers be so interested in these new technologies? Peer pressure. Not the kind you think of when you think of two teenagers trying to convince another to try a cigarette or gulp a beer. We are talking about the type of peer pressure that positions you in the hierarchy of your peers. Recall the importance of know-how and knowledge mentioned earlier in this chapter? Moms are using new technologies such as blogs and podcasts to illustrate and share their knowledge of parenting, business and social issues because it's an important part of building relationships with others. The more we know about motherhood, the more valued we are as moms in the eyes of our solution-seeking peers. Fortunately for marketers, there is room to inhabit in this pyramid of seek and supply.

Leveraging New Technologies

Companies can leverage the power of knowledge and know-how by empowering moms with more of what they need. New technologies are an effective way for marketers to do this in two different ways. First, you can connect with moms by supplying them on-the-go knowledge and know-how. The delivery of this information can come in the form of a typical podcast that relates to your product or a blog from your designers. It can be video content that explains how to assemble information or usage ideas. The content you offer can also relate to the lifestyle or solutions it provides. For instance, a crock pot maker might offer cooking videos online, complete with downloadable recipe cards. An office supply retailer might send out e-newsletters with tips for balancing work and family for female business owners. The key to your success is that the know-how or knowledge of what you are providing is relevant, credible and valuable to the lifestyle of your target mom.

The second way to utilize new technology is to create a platform that empowers moms to share knowledge with each other. In this case you are part of the conversation from the standpoint that your company has provided a powerful tool for sharing know-how. Your part of the conversation might be seen rather than heard; a quality that many moms appreciate with their children. An example of such an initiative is Lady Speed Stick's website, www.mylife247.com, in which moms can submit journal entries about how they juggle the demands of their day and cram so much into 24 hours. Lady Speed Stick created the platform which allows moms to share their experiences, with room to deliver their branding to mothers who visit the site. In this instance, moms themselves become a major player in Lady Speed Stick's marketing campaign.

It would be too easy for me to advise marketing professionals to simply design marketing programs that mimic the personality and behaviors of mothers and use new technologies like social networks and videos. However, it's far more beneficial and interesting to examine how mimicking the Mom Market can put you in the game of new media marketing and truly connect you to this lucrative consumer group. Although this book focuses on tactical approaches to using new media to reach moms, your marketing plan should be built on the foundation of a few key elements. These strategies include relationship building, community, and customization. What these three topics have in common is their tie to the five core values of a mom.

The Five Core Values

Interestingly what we found after pouring over thousands of pages of research, conducting hundreds of interviews and observing millions of moms globally, is that moms regardless of their age, race, ethnicity, family size or geographical location share five core values. They are: 1) Health and Safety; 2) Family Enrichment; 3) Value; 4) Simplification; and 5) Time Management. Moms use relationship building, community and customization to connect with these five goals. They are the channels that moms leverage to ensure her children grow up safe and healthy with the hope of having "more," whether its food or education, in the most value-minded simplicistic way possible within her limited amount of time.

It's one thing to engage her in your company's blog or have her viewing your videos, but it's another for her to say, "they get me!" The latter creates a deep bond between you and your target market. The only way to do this is to share her core values and work toward a common bond. Understanding the importance of relationship building, community and customization will maximize the effectiveness of your marketing.

Relationships

Let's start with one of the most important touch points of any mom: relationships.

Every good businessman or businesswoman knows the importance of relationships. Nurturing relationships speaks to the core of a mother's existence. From the first moment of life, she strives to bond with her child. Moms inherently nurture relationships. They nurture relationships with almost everything they come in contact with throughout the day. When they shower with Dove soap and run the soap over their less-than-perfect thighs or stomach bulge, they feel connected with the full-figured women that adorn Dove's billboards. This is one of the reasons the Dove Real Campaign resonated so well with mothers. It was a relationship based on trust between the brand and the woman watching it. She could trust that Dove would understand her, even with the extra ten pounds of baby weight still lingering on her buttocks. At CVS, she feels part of a union when she receives coupons for future purchases after showing her CVS card at the register. At Starbucks, she feels appreciation for the brand that allows her to get her coffee just the way she likes it. Finally, online she feels a commitment to Snapfish who not only organizes her digital photos, but also anticipates her needs to turn them into gifts and cards.

Not every point of contact ends in a positive relationship however. Just like children can disappoint mom with their behaviors, brands can miss the mark too. This happens when some marketers ignore the potential of establishing a relationship, while others fail to live up to their brand promise. Additionally a disconnect can happen when a company sends out confusing messages that make it difficult for busy moms to understand the scope of the relationship. Few market segments are as receptive to establishing a relationship with you as today's mother, which means a great upside to your sales goals.

Relationships Gone Bad

It's only fair to mention what can happen when the relationship goes bad with moms. If you thought that hell hath no fury like a woman scorned, then you don't want to cross a mom who has been let down by your brand. BSM Media research shows that a mom will tell up to 20 other mothers about a product or brand that has failed to live up their promise, or has treated her poorly. What is worse for you is that she will use the same technologies that she uses to spread her praise of your product to help her peers avoid the perils of your product. She will text, vlog, blog and post about the experience as a way of educating other moms. These mothers don't see it as bashing your brand; she will use the bad experience with your product to nurture her relationships with her peers. She sees it as helping other mothers save time and money. In the hierarchy of mothers, she will be elevated because she has saved other moms from wasting precious time on a product that doesn't work.

In the same situation, if your product exceeds her expectations, she will spread the news among her peers because the knowledge of great-finds elevates her again among other moms. This mother becomes the "go-to" person because she is in the know, whether the information she's passing on is positive or negative. In the world of moms, it all comes down to the fact that one mom knows something that her peer groups don't know yet, establishing her as the influencer. Information, knowledge and know-how are important currency among moms. You'll see how "knowledge and know-how" plays an important part of effectively communicating to this consumer group later. In the chapter dedicated to Mom Mavens we will examine how marketers can leverage the influence of mothers at the top of the peer pyramid.

Communication

Knowledge is only a small part of the larger tool moms use to nurture relationships. Communication is the overriding means of moms getting her message distributed and absorbed by the people around her. The strategy of utilizing new media speaks directly to her core definition of communication, if you consider it a multi-channel delivery system. For a marketer it might mean a two-way exchange that is delivered across online video, message boards, interactive blogs or customer generated content. For a mother communication can be a hug, a pat on the back, a screeching yell to clean up your room or an email filled with junior's birthday pictures. Basically, communication means that I am going to engage myself into the situation—by opening an email or picking up my phone or reading a webpage—and, someone is going to respond back.

If there is any doubt in your mind that women place a high price on the "speaking back" part of communication, let me give you an example that might hit home, particularly for male readers of this book. Have you ever gotten in trouble with your wife, sister, mom or significant female in your life because you didn't call when you were going to be late? Even if you decided it was quicker to get home rather than pick up the phone? And, after suffering the wrath of her emotions that ranged from concern to anger, you still don't understand why she is so angry?

Women like two-way communication. They expect you to anticipate their emotions, respond at their speed, and tell them when something changes in your relationship. Overlaying this is the emotional tie to appreciation and respect. No wonder men get in trouble for not calling. The lack of communication says to a woman, "I don't respect your time or appreciate the fact that you might be waiting for me." The same undefined rules of engagement apply to a mom's relationship as a consumer. She wants to know that there is a customer service operator she can call if she has a problem. She wants you as a company to respect her opinions and ideas, and she desires a way to complain when things go wrong. It's important that moms feel the relationship with your company is a two-way street. Leveraging the marketing techniques in this book will make it possible for you to maintain a meaningful and consistent dialogue with today's mother.

Suave does a very good job of creating a relationship with moms through their online destination, www.inthemotherhood.com. They successfully use video, message boards and consumer-generated content to engage moms in their brand. They create an experience that allows a mother to customize the relationship based on her preferred delivery channel. In actuality, she can write, read, vote for, or view her relationship with Suave. It says to a mother, "We understand you." Nothing could mean more to a mom than to know that someone else relates to her life, no matter how alienated she might feel lost in pages of never-ending to-do lists. We will examine Suave's marketing initiative more closely later in the book.

Now that you have seen how much moms value communication you can see why it's so crucial to implement new marketing techniques such as video, blogs and social communities to reach her. As a marketer you want the ability to establish and nurture a relationship with moms. In order to establish the relationship you must be able to interact with her along the path of her busy day. This requires you to speak to her wherever she is whenever she's there. It's not enough to be online if mom is in the carpool line at 2:00 p.m. You miss your target if you have all your advertising in television if she's online from 10:00 p.m. to 2:00 a.m. This is why it is important to utilize new technologies and new marketing techniques in order to maintain a dialogue with a fast moving target.

MomMe-Marketing

Until the late 90's women with children were viewed as a "niche" in the marketing world. They were considered a sub-segment of female consumers and although brand managers knew they were vital targets, they lumped moms in with the bigger target of women. Marketers chose to cast a large net into the pool of women and hoped to pull out a few moms in the process. Basically, marketing to mothers was a numbers game. Even in the dawn of the Internet advertisers hoping to connect with moms would do large media buys on female oriented sites searching for sales among mothers. Perhaps the diversity of the Mom Market seemed overwhelming to marketers and media buyers. After all, what group of 82 million individuals is comprised of such an intricate web of behaviors, beliefs, lifestyles and life stages? It's a lot easier to market to the common denominator of gender than it is to develop ad copy focused on marital status or parenting style.

Fortunately for us all, Starbucks began to allow mothers to customize their coffee and Build-A-Bear Workshop enabled her children to personalize their teddy bears forcing marketers to take a different approach and permitting consumers to expect more. Today customization and personalization is one of the most important strategies of connecting with moms. It allows marketers to develop a relationship with a mother that is relevant and valuable regardless of her career choices or parenting style—a very valuable tool when trying to connect with a market as diverse in lifestyles and life stages as the children they are parenting. Today customization and personalization has become a consumer expectation when interacting with products and services. While it may be a new strategy for marketers, mothers have been engaged in the practice for a very long time.

Let's take a look at products and how moms customize design to their lifestyle. For a long time, on-the-go moms have made mini bags of snacks and treats for her on-the-go children. A decade ago it was not uncommon in a classroom to find a mom distributing snack size baggies filled with crackers because she needed single-serving sizes when feeding groups of children. However it was not until 2001 that food manufacturers caught on to this idea and began producing single-size 100-calorie packs. Now the shelves are filled with products that moms have been manufacturing in their kitchens for years. Another good example of customizing products to their needs is paper towels. BSM Media conducted a survey of 400 mothers to determine how many uses a mom had for paper towels. These moms were customizing the use of papers towels in 536 different ways, using them as tools for everything from drawer liners to love notes in their child's lunchbox.

Moms are using customization and personalization when it comes to digital photography as well. Pictures have always been an important part of a mother's life. It's her way to chronicle the events that take place within her family, a way to share her children with family members, and it is increasingly becoming a platform in which she can personalize gifts. Once the photo is captured on her digital camera many things can happen. She can decide to use it as a screen saver, she might print it from a home printer, send it to a retailer who will print it for her, or store it with an online service. She may decide to create 4x6 prints, turn it into a card, incorporate it into a calendar, or print it on a coffee mug. The mom has so many options and so many opportunities to personalize the experience with that photo. She also has many ways to create her ecosystem surrounding digital photography and she is doing just that. Ask any mom about pictures and she will likely go into a list of services, retailers and tools she uses to store, organize and print her photos. There is the opportunity for technology providers, retailers, and online services to give her enough ways to customize the experience with their brand so that she will work them into her ecosystem.

For marketers, customization and personalization offers a format familiar to mothers and a means to create meaningful relationships. There are many ways to successfully utilize these important tools. Online, many marketers such as Snapfish allow mothers to personalize gifts by uploading pictures to their site. M & Ms enables moms to add personalize messages to their famous chocolate candies. This has not only increased sales but offered mothers new ways to consume and utilize M & Ms, from baby showers to birthday party favors. Creating a personalized experience with your brand can not only help you build sales but also build a community of moms who interact with you on a regular basis.

Community

Moms are part of multiple communities just as brands need to be a part of various communities of consumers in today's marketplace. They create networks of relationship within families, in business and in the community. Among themselves, they create book clubs, social selling groups such as Tupperware and Discovery Toys and playgroups. In order to fully engage with the Mom Market, marketers must become active in two types of communities. First, they must be active members within mom's physical community. This includes her playgroups, momprenuer business networks or organized sports teams. Secondly, companies must create their own type of peer-to-peer communities. This represents strategic partnerships that expand your reach and put your products and marketing messages in unexpected places. We will examine tactical ways for you to do this in later chapters. For now, let's examine the dynamics of mom communities in general.

In the world of moms, community is defined as an arena for socializing. Not socializing in terms of having a few drinks and shooting the breeze but rather a forum for sharing, learning and connecting. As you go through this chapter, it is important to remember that when we discuss socialization, we are speaking to the actions that mothers exhibit within a community. Moms are social beings by nature. Her social channels allow a mother to experience many emotions such as a sense of accomplishment, pride, connection and friendly competition. For example, a mother of a preschooler may belong to a playgroup in order to allow her child to interact with other children of the same age. However for the mom, it is an opportunity to demonstrate her parenting skills among other moms, benchmark the progress of her child against other preschoolers and build her network of peers.

Social events allow moms to brag in a subtle manner. For instance, a simple birthday party might become a statement of skill when a mother hand-makes the invitations or customizes the party favors for each guest. Friendly competition is as much part of socializing as nurturing relationships with peers. Today's moms socialize in many different ways. It's no longer just getting the kids together at the park or meeting up at the baseball diamond. Moms socialize in businesses, online, and in the community through podcasts and via wireless devices.

Most of us probably hear of moms getting together in communities and think of playgroups, PTOs (formerly known as PTAs) and organized athletics. However moms are mixing socializing with business. For instance, mothers have created a huge community in the direct sales industry. There is a large number of momprenuers who serve as distributors, hosts or representatives of direct sales companies. This didn't happen by accident or at the recruitment of any one company, but rather moms see direct sales companies as a solution that fits well into her ecosystem of work/life balance.

Social selling is the perfect fit for today's mothers. What do direct sales companies such as Tupperware, Discovery Toys, Northern Candles and Pampered Chef offer to moms? They deliver an opportunity to earn extra income while allowing mom to get out of the house a few hours a week to interact with her peers in the convenience of another mom's living room. It takes girlfriend groups to a new dimension. Direct sales enable a mother to personalize her work with flexibility while sharing information and knowledge with other moms, two very important touch points for a mom. With each party, a new community of moms forms.

Online moms are active in virtual communities. The strongest groups are ones in which members embrace a passion for a common theme. For instance, the virtual community of BlueSuitMom.com, has over 500,000 loyal moms because each of these women feels a connection to the challenges they face in balancing work and family. BlueSuitMom.com gives them an online community to share their challenges and solutions. Home-based Working Moms, www.hbwm.com, is an online community of moms dedicated to being home with their children yet they share a desire to run a business to facilitate their career choice. There's even an online community for moms who enjoy running at www.seemommyrun.com. Here over 5,000 moms share stories of their passion for running even though many hit the pavement thousands of miles away from each other. And finally, Amazingmoms.com is a virtual community for moms with a passion for throwing great birthday parties.

On these sites, moms share ideas, pass along tips, trade product information and develop friendships. The virtual relations between some of these women date back to the mid-nineties and the bonds are as strong as sisterhood. Today moms are using new technologies to increase the membership of their communities, the frequency of contact and to enhance their ability to share. They are also using these technologies to customize and control their communities. On Newbaby.com, moms can upload video of their children and share it either with all moms who visit their site or with select moms who they choose to be in their profile group. Ironically as technology allows moms to increase the reach of their communities, moms are becoming more selective as to the size, style and format of their online communities. Essentially they are applying their desire for customization and personalization to their communities.

The Touch Point

In order to develop a meaningful relationship with moms as we discussed earlier, it's important to be a part of their community. But what role can a company play in a girlfriend group that focuses on infertility or home schooling, particularly if the company is a food manufacturer? There within lies the challenge.

Marketers must determine a touch point that is relevant to moms and present it in a way that is neither commercial nor patronizing. Don't expect that you are going to be a girlfriend in their girlfriend group just because you make it into the room. Brands must give these women something of value and relevance as well as a sense that you validate their purpose.

Here's an example. Over the past decade, my team at BSM Media has created The Guerrilla Mom Network, a group of middle to upper level mom sites with loyal virtual communities. Instead of placing banners on some of the larger parenting sites who have audiences of 1 million plus, we choose to distribute content within the Guerrilla

Mom Network. The strategy behind these campaigns is that many Mom websites are produced by passionate moms who are self-funding their desire to deliver information to other moms. The burden of writing fresh content often falls on their shoulders and while customer generated content has made it easier to keep their sites updated, they still seek new articles.

We produce several pieces of content that integrates the company's product message into it and pay for placement of the article on these sites. Almost 100% of the time, these mom webmasters over-deliver in number of impressions and the response rate by audience is higher than when we place banners on big mom sites. The explanation is interesting. First, the mom webmaster is so thrilled that a big respected brand is spending money on their site that they over deliver on impressions. Second, money means validation so if a company spends money on a small community site; it validates the purpose of the moms who are members of the site. Moms reward this validation with click thrus and ultimately by buying your product. The loyalty you experience as an advertiser is an outward sign of a mother's desire to nurture relationships. It is "You support me and I'll support you" in living color. And because moms are social, they tell others within their community to support you. You've officially earned your spot in their community.

There are other ways to enter a mom's community online. It goes back to the power of knowledge and practical know-how. A brand can become an active member of a community by supplying moms with information that they feel is valuable and applicable to their life. As member of a community, you are expected to develop a dialogue and your conversation can start with content. Think about solutions that your product produces for mothers or themes that surround the use of your product. This is excellent content for the dialogue you want to maintain within these communities.

For instance, if you are a car manufacturer and through your knowledge of moms you know that 70% of children do their homework outside the home, you might provide a piece of content to a mom site that caters to school age children and is titled "Ten tips for managing homework on the go." It's relevant, it's helpful, and most importantly it says to mothers, "We know what happens in your vehicles."

The majority of this book will be focused on discussing the tactics for delivering content. It doesn't have to just be an article anymore and can take the form of a Vcast, Podcast or blog and it doesn't have to come directly from you. One of the most effective delivery mechanisms for companies is to have another mother deliver your marketing message. I've devoted an entire chapter to recruiting, engaging and using Mom Mavens and Influencers to be your brand evangelists. First, however, let's examine the women that comprise today's Mom Market.

Moms are chatting, tweeting and blogging with friends, family and strangers. One thing they all have in common is that they are time-starved.

Who Is Today's Mom?

CHAPTER

IT HAS BEEN ALMOST A DECADE SINCE I PUBLISHED MY FIRST BOOK ON the Mom Market. It was 1999 and no one, not even the US government, had quantified the impact of mothers on the US economy. I recall sitting in our office with my business partner at the time, Rachael Bender, and shifting through pages and pages of data to calculate the annual spending of US mothers. It literally took us an entire day of numbers to get to the $1.7 trillion that is often quoted in the media and even the websites of marketers who have hung out a Marketing to Mom agency shingle. Every now and then a savvy research analyst or tough editor will ask me how it was calculated. The short answer is that the annual spending of mothers is determined through data from the US Census Report and the Department of Agriculture, which is the agency that tracks spending. Overlaid on this calculation is the amount of household spending a mother controls.

Here's where it starts getting more complicated. In order to come to the most accurate conclusion, you must consider the structure of the household. Single mothers control almost 98% of the household income while married women control 85%. The greatest change in variables since we began calculating Mom Spending is the increased number of single moms. In fact, according to the US Census 2000 Report, single moms are the fastest growing segment of the US population and currently number over 10 million. You can see as this number increases, so does the annual spending of mothers. This is not the only fact contributing to the increase. Mothers are increasingly starting businesses, involved in eldercare and influencing the larger purchases in their household. Women are also becoming mothers at younger and older ages therefore widening the pool of population that falls into the category of "female with at least one child living in the home." Just a side note, we do not calculate women under 18 with a child because we feel that their spending influence may be affected by other supporting factors.

Every year, my team recalculates the data to determine the current spending of the 82 million women with children in the US. As you can imagine, the number climbs year over year. Today, mothers in the US control over $2.1 trillion dollars a year. Single mothers alone account for $174 billion in annual spending.[1] As I often say, $2.1 trillion is more than the gross national product of Portugal, Australia and Spain. It's a lot of money!

What this represents is a utopia for companies who have a product, service or information source for moms, which in my mind is just about everyone. If you are taking the time to read this book, then you probably don't need to be convinced about the spending power of mothers. Which, by the way, is a far cry from where I was in 1999 when 90% of my time was spent convincing companies, from Coke to Burger King to Office Depot, that moms were lucrative customers

1 Young & Rubicam, "The Intelligence Factory," *The Single Female Consumer* (Young & Rubicam, July 2000), www.yr.com

and it was not enough to market to women to tap the Mom Market. I'll assume you are here right now because you understand the value of moms and want to connect with them through new media and new technologies. For this reason, I will only devote a small portion of this chapter to the actual numbers behind their spending power and more on the words to help you understand the personality, behaviors and characteristics of today's mother.

The Mom Spend

Moms have emerged as consumers. Companies once saw them as purchasers of diapers and groceries. Today mothers are not only controlling household spending, they are spending money in business as well. Women, many of whom are mothers, are starting an estimated 4 million businesses a year.[2] They are buying office supplies, copiers, fax machines, computers, and accounting and legal services. Mothers, looking for a way to balance a career and family, are starting home-based businesses at four times the rate of their male counterparts.[3] These companies range from small one-person service providers to multimillion-dollar businesses. Women employed outside the home control $1.5 trillion in spending for business.[4] This spending is coming not just from mothers in business for themselves but also from mothers employed by others.

A Census Bureau report cites that even among mothers with very young children, more than 60% are in the labor force.[5] A typical middle manager will decide where her department goes for a celebration luncheon, or how to budget advertising expenses, or whether to send a colleague to an out-of-town meeting.

Moms represent not only power in the economy of the US, but in

2 Center for Women's Business Research, *Women-Owned Business in the United States, 2002: A Fact Sheet* (Washington, D.C.: Center for Women's Business Research, 2002), www.nfwbo.org.

3 Ibid

4 Connie Glaser, "The Women's Market Rules," *Competitive Edge* (May/June 2001).

5 U.S. Bureau of Labor Statistics, *Working in the Twenty-first Century* (Washington, D.C., 2000).

politics as well. Mothers represent an important block of votes for candidates, particularly those running for president. The so-called soccer mom contingent helped put Bill Clinton in the White House in 1996. In the last four presidential elections, suburban mothers have represented one-fifth of all the votes cast.[6] In fact, I recently had the opportunity to have an informal conversation with Elizabeth Edwards, wife of former presidential candidate John Edwards, and she agreed that US mothers represent one of the most untapped pool of voters today.

Mom's Money

- According to the U.S. Department of Agriculture (the agency that tracks family expenditures), an average-income family will spend $165,630 on a child by the time the child reaches eighteen years of age.[7]

- Parents with incomes of $38,000 to $64,000 spent $18,510 on miscellaneous items for the average child from birth through the age of eighteen. This includes spending on entertainment, reading material, VCRs, summer camps, and lessons.[8]

- Females outnumber males in the United States by over six million (roughly 6 percent) and a significant percent have at least one child. There are 141,606,000 women with children in the United States.[9]

6 James Bennet, "Soccer Mom 2000," *New York Times* (April 9, 2002).

7 U.S. Department of Agriculture, *Expenditures on Children by Families*, Washington, D.C.: GPO, June 2001.

8 Ibid

9 Bureau of the Census, Population Projections of the United States by Age, Sex, Race and Hispanic Origin: 1995 to 2050, Washington, D.C.: GPO, 1996 Business Research, Women-Owned Business in the United States, 2002: A Fact Sheet.

Mom's Money

- Woman-owned businesses generate $1.15 trillion in sales.[10]
- Women-owned businesses employ 9.2 million people.[11]
- U.S. women spend more than $3.7 trillion annually on consumer goods and services, plus another $1.5 trillion as purchasing agents for businesses. As a group U.S. women constitute the number three market in the world, with their collective buying power exceeding the economy of Japan.[12]
- According to BabyCenter, www.babycenters.com, the average child will cost $338,000 by the time they finish a public college. Before they are 18 years old, housing will cost $105,000, food will cost $41,400 and health care will cost $17,400.[13]
- Eighty percent of all checks written in the United States are signed by women.[14]

10 Center for Women's Business Research, *Women-Owned Business in the United States, 2002: A Fact Sheet* (Washington, D.C.: Center for Women's Business Research, 2002), www.nfwbo.org.

11 Ibid

12 Connie Glaser, "The Women's Market Rules," *Competitive Edge* (May/June 2001).

13 Faith Popcorn, *EVEolution: Understand Women: Eight Essential Truths That Work in Your Business and Life*, (Dimensions, 2001.)

14 Faith Popcorn, EVEolution: Understand Women: Eight Essential Truths That Work in Your Business and Life, (Dimensions, 2001.)

Here's where we will end the discussion on numbers as it relates to mom's spending. It's a fair conclusion that the amount of money mothers control in the US will continue to grow, at least for the next decade and marketers will want to tap this lucrative market. In order to accomplish this it's important to truly know your target market. This is the goal of this chapter—to put you into the heads of today's mothers, to help you walk in her shoes and ultimately put you into her purchasing patterns. Let's walk in her shoes and into her day.

The Mom's Day

Moms are doing more than just spending money. They are engaged in everything from washing clothes to budgeting business plans. They are online, offline and off-road. Moms are chatting, tweeting and blogging with friends, family and strangers. One thing they all have in common is that they are time-starved. According to BSM Media research 64% of moms say they have given up sleep in order to get work/chores done so that they could spend time with their children. The role of mother can be isolating. In fact, 74% of moms surveyed admitted feeling isolated from family and friends after giving birth. Perhaps this is why over 40% of them chat and socialize with other moms online. While online, moms are sharing information with others and empowering themselves to do more with less time. About 30% have blogged with 66% having posted a message on a board or chatted online.

Moms share a lot. Sixty-four percent of mothers ask another mom for advice before a large purchase. They also share memories. Moms are the chief archiver, taking 80% of the family photos. They like to customize products and have fueled the growth of customizable brands such as Build-A-Bear Workshop, M&M.com (customizable candy) and American Girl Dolls.

Moms are efficient. In fact, 55% indicate they have Halloween planned prior to Labor Day and 65% have thought about their Fall calendar by the first day of school.

I've sliced and diced moms in many ways over the years. In *Marketing to Moms: Getting Your Share of the Trillion Dollar Market* I focused on moms in general. In 2005 my focus turned to examining the three generation of mothers in *Trillion Dollar Moms: Marketing to a New Generation of Mothers.* Today many of the core values and motivators remain the same for moms while their consumer behaviors and media engagement have changed. I use the word engagement here rather than consumption because today's mothers are using media and technology to engage in marketing, shopping and dialogues with companies. In fact, if my first book was written today, it would be called "Marketing WITH Moms" rather than "Marketing to Moms." The following chapters will fully illustrate how the most effective marketing programs are joint ventures with today's mothers, rather than strategies, to deliver brand messages to moms.

When I wrote *Trillion Dollar Moms*, I did so because I had successfully convinced marketers to focus on the Mom Market. However when they did so, they realized that there were so many types of mothers in the market that they didn't know who to target. There had been a societal shift in the mom population that even I had not fully noticed. Women from three distinct generations were, for the first time in history, becoming mothers. Older Boomer moms were giving birth in hospital rooms right next door to twenty-year-old Millennials and I was giving birth to a new book that examined how these generations of mothers were alike yet different. This historic make-up of mothers still exists today, and for that reason I believe a condensed overview is beneficial for marketers trying to tap the Mom Market.

Historical Make-Up of Mothers

The U.S. Census describes Baby Boomers as people born between 1946 and 1964. Of them, approximately 40 million are mothers. Generation Xers were born between 1965 and 1980 and account for 15.6 million mothers. Generation Y, or Millennials as some call them, were born between 1981 and 1995, and although they currently represent the smallest pool of mothers, they are expected to increase the current number of 4 million births a year as they begin to become mothers.

Generation Y	**1981-1995**	**57 million**
Generation X	**1965-1980**	**50 million**
Baby Boomers	**1946-1965**	**70 million**
Silver Birds	**1934-1945**	

Boomers Moms

Mothers of this generation no longer function within the boundaries of generational characteristics as their predecessors did. Boomer moms are less likely to act their age, instead taking on the behaviors of the cohort that most resembles the moms of their children. For instance, a forty-five year-old Boomer mother who has a toddler will behave more similarly to Gen X moms who have toddlers. This is a critical bit of information for marketers who have been segmenting mothers by their age rather than the age of the children. Our research proves that the age of the child is a far more critical segmenting tool than the age of the mother. This was the big takeaway in *Trillion Dollar Moms* and the behavior still exists today.

Boomer moms are the group dubbed "soccer moms." They have an elevated sense of style and are willing to pay more for the items they value. They will purchase their children's clothing at Target yet buy their shoes at Nordstrom's. Their disposable income is slightly higher than the two generations of mothers that follow because they are further along in their life stage.

Nostalgic marketing works well with Boomers because they seek opportunities to recreate their own childhood memories for their children. Many of these women have blazed the trail between work and family and because of this it takes a village to raise their children. They feel extremely time starved as they struggle for balance in their lives. This is why recreating memories is important to them. If they can rely on brands to leave the same pleasant memories in the minds of their children then it is one less challenge for them to maneuver. For instance, if a Boomer is shopping for crayons and she is standing in front of Crayola and Rose Art brands, she will select the Crayola crayons because she remembers the scent of fresh Crayolas on the first day of school. She feels assured that the first day of school will be as special for her own child thanks to Crayola. "Good first day" is now off the overcrowded task-list for this mom.

Family enrichment is very important because she is always trying to squeeze in quality time with the family. Marketers can leverage her values by not only creating memories but creating experiences for her to enjoy with her family. Boomers with older children are now becoming empty-nesters. With their newly found freedom they are starting businesses and discovering new hobbies without relinquishing their role as a mother.

Generation X Mothers

These women have grown up with technology in all forms and multimedia advertising with more options for receiving and seeking information than all previous generations combined. As mothers, they expect marketers to appeal to their multi-sensory communication behaviors. They will challenge marketers to think well beyond the typical marketing toolbox to deliver messages in new ways that integrate into their lifestyle.

Generation X moms are currently 20 to 39 years old and identify the group born between 1965 and the early 80's. They number approximately 51 million and as a whole, Xers represent approximately 25% of the population.[15] Generation X is particularly difficult to generalize because it is the most diverse generation in history. It is made up of more ethnic backgrounds than any of its predecessors. Generation X is the largest population of naturalized citizens in U.S. history.[16] Thanks to negative and often unfounded media coverage that depicted them as a generation that was lazy, whiny, cynical and disloyal, one common theme exists within this group of men and women. They hate to be called Xers.

Generation Xers are also known as the generation of "latchkey kids," a description for children of working mothers who symbolically carried keys around their necks so they could let themselves in after school. These children had to become self-reliant and fend for themselves while their Boomer mothers worked long hours away from home. The outcome of long hours alone produced a generation that grew up confident in their problem-solving abilities.

One of the greatest influences of this generation is divorce. This was the first generation since the Beaver Cleaver household of the 50's to have to learn to manage divorced parents, splitting their time between two households and developing relationships with blended families. Some experts estimated that nearly half of Gen Xers grew up in divorced households based on the Census Bureau number that 50% of marriages in the late 60's and 70's ended in divorce.

One effect of the uncertainty they witnessed among their parents was a lack of desire to get married. According to the 2000 Census, single mothers comprise the fastest growing demographic in the US.[17] Surprisingly, this group is not growing as the outcome of increased divorce but rather out of new moms' decisions to postpone marriage, regardless of unplanned pregnancies. In

15 U.S. Bureau of the Census, *Population Projections of the United States by Age, Sex, and Race*: 1995 to 2050 (Washington, D.C.: GPO, 2000), www.census.gov.

16 Ibid

17 U.S. Bureau of the Census, *Population Projections of the United States by Age, Sex, and Race*: 1995 to 2050 (Washington, D.C.: GPO, 2000), www.census.gov.

addition, the freedom to stay single is made easier by the Generation X mom's ability to financially support her family on her own.

Although most moms will solicit the advice of their own mothers after giving birth for the first time, the metamorphosis of the relationship takes place as it moves from simply mother-daughter interaction to a friendship based on the common experience of motherhood, and the mother's advice suddenly carries double weight. However, it should be noted that because of the nature of a Gen Xer to be self-reliant and create their own solutions, they would ultimately make their own decision based on what's right for them personally.

When a Gen X mother is seeking advice on baby food, she will turn to her friends or members of a peer group and ask them to share their personal stories and experiences with various brands. Gen X moms like stories because they illustrate the individuality of different situations. They allow her to see that there is more than one way to do things. Once she gets insights from peers, she'll test this information with her mother. Not surprisingly, the Baby Boomer grandmother will tell her how things were when she was a child and what brands she used to feed her babies as a young mother. The Gen X mom will discount stories of the "old days" in an attempt to forget her own unstable childhood and avoid recreating it for her child. What changes the scenario is that the Gen X mom has now repositioned her mother as a friend rather than just her mom and will extrapolate the valuable facts from her mother's advice.

Generation Xers are on a quest for individuality. They feel that being their own person allows them to better control the uncertainties of the world they experienced as a child. The quest for individuality is seen on retail shelves with the popularity of personalized jewelry, sweaters, stationery and household items. It is also seen in the number of customized names for moms. From "Yoga moms" to

"Slacker Moms" Generation X mothers are maintaining more of their own personal identity and adapting motherhood to fit with who they are as women.

A stable home-life is the dream of many Generation Xers—particularly those who grew up either as a latchkey child or as the product of divorce. It's been said that everything eventually comes back into style and this may be true also for the traditional nuclear family. The Cleaver family may come to mind, but I ask you to leave your preconceived visions of mom aside. Today's Generation X mom desires to create a stable home life for herself and her children with or without a man. While Baby Boomer moms believed it took a village to raise a child, Generation X moms believe it takes a family.

Generation Xers desire and have high expectations for groups. They like to form and belong to groups where they can share ideas and find opportunities for growth, creating surrogate families of like-minded peers who share their vision of community and chosen family. With unlimited groups that exist for just about every interest, personal value or belief, whether in person or virtually, Gen Xers can express their individuality and find common ground with others across the city or across the country. Groupings by Xers have spawned a myriad of new social interactions among all generations. They include Mommy and Me playgroups, Bible study groups, volunteering events and home parties. In fact, the home party plan business has grown to a $29.95 billion industry.[18]

The influence of education is perhaps one of the most notable parenting priorities among Generation X mothers. Gen X moms prefer to spend money on enriching experiences that will create family memories like unique vacations and adventure travel. It's not enough to buy airline tickets and nights in a posh resort.

18 *2004 Direct Selling Growth and Outlook Survey Fact Sheet*, http://www.dsa.org (accessed September 14, 2004).

Gen Xers are active consumers. They weigh their buying decisions against the personal and financial costs they set for themselves. In other words, before purchasing a product or service, they will decide if it's worth it for them in the investment of time and/or dollars. They are very realistic and define value against their sense of realism. They can't be sold and see marketing at face value, looking past flash and sizzle to make purchases based on their own needs. They want results and benefits, plain and simple. Their pragmatic attitude on life comes out in what they expect from marketers. They want real answers to real life challenges. And this grounding in reality and the desire to speak the truth has birthed the new genre of television entertainment, reality television, and the increased interest in live broadcasts, eliminating the ability of someone behind the scenes to manipulate the story or outcome.

Gen Xers are visual in their learning, a trait acquired while watching TV and exploring the early years of computer games and activities. They enjoy imagery but recognize that it is just an outcome of technology. Their ability to juggle tasks, a skill they learned as latch-key kids, enables them to absorb multi media messages at one time. They have the mental capacity to take in and process images that are being delivered to them through multiple channels. Gen X moms tend to be traditionalists, perhaps as a way to create stability for their children. The resurrection of nostalgic toys, games, and retro designs has in a large part been fueled by the Gen Xer's desire to hold onto a conservative traditional family life.

Generation Y

Born between 1983 and 1994, they are most often defined by the fact that they graduated from high school in the new millennium. In 2004, they are aged 7 to 25 years old. They number 70 million and make up 21% of the U.S. population.[19] For marketers, Generation Y

19 U.S. Bureau of the Census, *Population Projections of the United States by Age, Sex, and Race*: 1995 to 2050 (Washington, D.C.: GPO, 2000), www.census.gov.

presents the largest consumer group to emerge since the Baby Boomer generation. It is estimated that the population will grow by twice the average rate as Generation Y continues to grow until 2010. As a consumer group they control approximately $172 billion a year, and influence $300 billion, in family spending annually.[20] Gen Y has enjoyed a prosperous childhood. Over two-thirds of teenagers have a television in their room and when they aren't absorbing media visually, 70% of them are online. They are the most socioeconomic and ethnically diverse population in the history of the US. Minorities make up 34% of Generation Y, up from 24% in the Baby Boomer cohort.[21] It is interesting to note that by 2010 Hispanics will be the largest minority group in this generation. They are a well-blended generation that celebrates diversity and is, in general, supportive of change that furthers inclusiveness and equality, such as acceptance of gay marriage.

Gen Ys have a passion for creating a better place to live, and an understanding of the state of the world beyond their front yards. Although three out of four Gen Ys have a working mother, their mothers are quite different from those that raised the latchkey children of the prior generation.[22] These mothers turned their attention to nurturing their young rather than chasing material possessions and career titles. These are the mothers who enjoyed movies depicting nurturing parents who made childrearing the focus of their lives and then ended up enriched by the experience, such as *Three Men and a Baby*, and *Baby Boom*. They have a sincere interest in parenting in a style we describe as "get on the floor and play." Self-confidence is one of the most defining traits of this generation, producing a spirit of ambition and passion that is

20 Zell Center for Risk Research Conference Series, *The Riskof Misreading Generation Y: The Need for New Marketing Strategies*, January 25, 2002, www.kellogg.nwu.edu/research/risk/archive.htm (accessed September 14, 2005).

21 Zell Center for Risk Research Conference Center

22 Ellen Newborne and Kathleen Kerwin, "Generation Y," Business Week Online, February 19, 1999, (accessed May 12, 2004)

illustrated in numerous areas. They are also aware of the damage done by previous generations to their world, and have grown up in a more environmentally sensitive climate than any previous generations. Self-assured Gen Ys believe they can change the world and correct the problems caused by the less-capable generations that came before them.

Technology has played a major role in shaping all aspects of the lives of Generation Ys. In fact, this was the first generation born into technology. Perhaps one of the most formative uses of technology is in socializing with friends. Communicating with peers has long been a popular activity with teens, but while Boomers spent hours telephoning girlfriends and Gen Xers later used email, instant messaging is today's technology of choice. In fact, two-thirds of teenagers use instant messaging.[23]

Gen Ys move and operate in groups, whether physically or virtually. The constant access these young people have to the relationships they value has instilled a sense of strong loyalty among their cohorts. They will be good mothers who demonstrate a great deal of confidence in this role. The practice they've had absorbing all forms of media at once has equipped them with multi-tasking skills like no other generation before them. Since they view education as a lifelong process, they will most likely instill the same values in their children.

When Generation Ys make buying decisions, they draw on their mother's message to them as a child: they can be whatever they want to be. Today they have plenty of images to try on themselves, from Britney to Eminem to Tiger Woods. They process what they hear and see in advertising messages, decide what they want to be and what fits their lifestyle and select product that meets their needs. They don't want to know how long a product is going to last or how it performed in taste tests.

---

23 Ibid

Three Generations Coming Together

Today's moms have evolved to a new place; what hasn't changed over the decade that I've been studying on them however are their core values. In fact, the core values we are about to discuss can even be called universal in scope. A mother in Japan places the same value on the health and safety of her child as the mom in South Africa or South Georgia. What changed is how they exhibit the value and the definition it carries in their culture. It's important to understand her core values for two reasons. First, it provides a shared goal in establishing a meaningful dialogue. Think about your closest friends. They've earned their spot in your circle of peers probably because you share a belief system, special interests or common priorities. It's the same criteria that mothers apply in developing relationships with brands, companies and service providers. If you can speak to one of her core goals, you will go a long way in developing a meaningful relationship with her.

The second reason it is important to understand her core values is that these values provide the foundation of motivations. You will see later in this chapter how her desire to simplify her life leads to her motivation to keep lists and find simpler solutions to everyday tasks. A mother's core values should be the starting point for every campaign, product design or engagement you create with, or for, a mother. Again, the five core values of mothers are: Health and Safety, Value, Time, Child Enrichment and Simplicity.

Health and Safety

Health and family safety rank at the top of a mother's core value list. Nothing means more to her than keeping her family safe and raising healthy children. I recently sat in the dining hall of Disneyland Paris. I watched moms from around the world feed their children breakfast. Now to the normal person, this would not be an elective activity at Disneyland; however for me it was a melting pot of insights. I was determined to find a "common" behavior within this group of mothers if one existed. Much to my family's dismay, I observed families for hours leaving the buffet line and sitting down to dine. It didn't take long however to find my common denominator. Unscientifically, I can tell you that 90% of the moms I watched made sure their children had food and were eating before they began to eat themselves. Call it a primeval behavior, but not one mom sat down without assuring herself that the health of her child would be maintained by a good breakfast. It's a very strong value—many moms will put their children's needs and desires above their own—and they will stay true to it.

Moms are smart enough to know that they alone can't keep their children safe and healthy so they turned to companies for help. They want to know that companies sincerely care about their health, along with their well-being. There are many ways a company can leverage this core value with moms. Companies who design products that tell moms they understand and care about her healthcare needs demonstrate their shared value through actions. It's not enough to just tout, "it's healthy." That approach doesn't work when a mom says it to her broccoli-hating child, and it won't work in marketing. Moms want a partner. They want someone to educate them and create a platform for discussion, for example in the form of a blog. They want to feel empowered by using your product.

Saving Time

Saving time might also be termed convenience. However you choose to phrase it, moms value time. Sometimes they even value it more than money. Whether a woman has one child or ten children, every mother believes they are starved for time. In the eyes of a mother, convenience is synonymous with saving time. Time for her family, time for herself, time to work, time for her spouse, and time for the house are just a few demands on a mother's time. The one type of time we don't want to talk to moms about anymore is "quality time," a loosely defined term that carries almost no meaning to mothers today. The term was overused in the mid to late 90's, and many mothers resent the term because it represents something not easily obtained and even less manageable.

Time is so precious that some mothers are willing to pay to gain a few seconds in their day. This is good for advertisers. Mothers' desire to find time can present a marketer with many communication channels. You can talk to mothers about how your product is going to give them more time, but another approach is to focus on what the mothers will do with the free time they gain by using your product. One of the most effective campaign slogans I've seen in a while was for a bank touting its convenience. They promised that by banking with them, the customer would regain two hours in their week. However it was not the two hours that was the motivator in the ad, it was this line, "We will give you two hours, what you do with them is your choice." It was brilliant because what one mother would do with two hours is quite different from what another would do. The campaign allowed mothers to place their own perceived value on the hours saved rather the bank doing it for them.

Value

Value is a word that moms understand well, particularly as they try to manage their family's finances. However one of the greatest mistakes marketers and retailers make in relation to leveraging this value is confusing it with price. Value doesn't always mean price in the mind of moms. It is more closely aligned with quality. To determine the value of a product or service, mothers measure its quality, convenience and relevance against the price. If she knows that paying a little more for higher quality will ultimately save her in time that she won't have to spend replacing the original product when it fails her, it's worth the price. The value of the product is greater than the price. Mothers don't necessarily want the cheapest product but they want to know that they are paying a competitive price and that the value comes in good quality, customer service and added benefits to her family. Here's where it all comes full circle. The benefits to her family will be measured against her core values. The more the benefits touch one of her core values, the greater the value it is assigned in her mind.

Child Enrichment

Child enrichment speaks to the natural desire for mothers to breed a superior offspring. It's the desire to do the job of mothering better than her own mother or provide her children with more than she had as a child herself. The trend is best illustrated within the Boomer segment of mothers. In the late 80's and early 90's these women accumulated material possessions like BMWs and Rolex watches as an outward sign of their success. Today, the children of this same generation have become the visual signs of success. It has produced a generation of over-scheduled children who are chauffeured from tennis to piano lessons, and private schools that have newborns on waiting lists for admissions. Messages that speak to mothers about bettering the lives of their children, enriching their experiences, and creating more intelligent students can be seen in print and electronic ads. The success of Baby Einstein videos was based on mothers seeking a way to better stimulate young, developing brains with music and exposure to the arts. Today's moms want their children to be smart and successful, but they also want them to be happy.

It's one thing to want more for your child; however too much of a good thing is never beneficial. Today many mothers have turned the value of child enrichment into an engagement in entitlement. These moms believe their children should be entitled to better things in life and their offspring have come to expect it. This is one trend marketers have seized with great zest. Spa parties for six-year-olds, mini boutiques for toddlers and preteen pedicures are among only a few of the initiatives created to capture the large population of entitlement minded mothers today. Although sales in these categories are growing, I am beginning to see a turn in the tide. In a recent BSM Media survey of 300 mothers, 80% of them said they would like to keep their little girls little for a longer period of time and saw a direct link to these pint-size luxuries as a contributing factor to accelerating their maturity. As moms go green, the economy weakens and the desire to simplify expands, I feel there will be a turn toward a more educational, philanthropic and spiritual based means of entitlement rather than materialistic ones.

The outcome to a mom's focus on child enrichment has a great deal to do with her desire to raise a happy, healthy adult. In fact when we ask moms what it is that she wants most from her journey in motherhood, that desire tops the list. Surprisingly it's not the smartest, wealthiest, or cutest adult when she gets right down to the answer. It's giving her child the tools, knowledge and experiences to become a happy adult. It's a job well done moment for her when she's accomplished this. Fortunately it's a long journey with many opportunities for marketers to be her partner on the path of parenthood with products and services to help along the way. It's important for marketers to also realize that the memories created along the way hold a significant value to her as well. She knows that when the kids are grown, memories will be all she has, and all her child carries away with them. This is why I elected to have a chapter on digital photography in this book. Photographs play a major role in chronicling the journey for mothers and their families.

Advertisers can leverage a mother's desire for family enrichment by speaking to the memories their product can create for the family. This is where nostalgic marketing emerges. Advertisers take mothers down memory lane in an attempt to help them relive childhood experiences that can be re-created for their own children. The message plays well to a mother's desire to create cherished memories for her own children.

A number of brands have recently jumped on the bandwagon of nostalgic marketing in order to appeal to mothers. For example, in the recent "More Ovaltine Please" ads, the message is the same as the one today's mothers saw when they were children. This reuse is effective because it allows mothers to relive their pleasant childhoods while providing a solution for getting their children to drink something nutritious.

Simplicity

The search for balance includes simplifying one's life, growing spiritually, and just feeling good. Marketers have done a good job in leveraging this value. It can be seen in magazines, merchandising, and of course advertising. Most obviously you can see it in the popularity of the magazine *Real Simple.* Take a stroll down the aisles of your neighborhood Target store and you'll find hundreds of products devoted to the idea of simplifying. You'll see shelf, closet, and home office organizers, color-coded toy bins, and seasonal celebration kits complete with holiday plates, cups, and napkins. Everything in one store with one mission: to make life easier. According to a study by the Radcliffe Public Policy Center, 61% of Americans would be willing to trade money for family time by giving up some of their pay for more time with their children or other family members. The value of simplification is that it produces time, the most important currency a mother possesses.

Values = Motivations

It may seem logical to you that a look into the values of moms is important in shaping the dialogue you establish with her. However the impact of her values can be felt way beyond establishing a common ground. Understanding her values allows us to understand the "why" behind many of her actions. Her values form the foundation for the 5 Key Motivators that ultimately govern and direct her actions. These Motivators are: 1) Nurturing Relationships; 2) Sharing; 3) Do it Simpler; 4) Healthy, Happy Adults; and 5) Accomplishment.

In her quest to raise healthy, happy adults, moms need to nurture relationships. These are relationships not only between her and her child but between her children and extended family members, peers, teachers and the community at large. She values time so she seeks ways to do tasks more efficiently in a simpler manner. This motivator will drive her to buy a new cleaning product or bundle products to create her own solutions. Mothers want to feel that they are enriching their family through their actions so they seek the sense of accomplishment. This quest for success motivates her to find the best vacation destination for each of her family members or build a home-based business that allows her to stay in the home.

As a marketer you want to touch her heart with one of her core values while leveraging one of her key motivators to produce an outcome that delivers the emotional fulfillment of reaching her goal. Let me give you one more example. If I were a cell phone company attempting to sell mom a wireless device for her child, I would speak to her value of safety while demonstrating to her how the wireless device can facilitate her ability to more simply stay connected with her child. Safety is what she values; however nurturing relationships by staying connected is what motivates her to purchase the phone. Fortunately these motivators run true for most mothers in today's Mom Market. But before you get too comfortable in your knowledge of moms, it's time to introduce you to the emerging sub-segments of mothers today.

Sub-Segments of Today's Moms

Not all moms are created equal and today's technology has made it easier than ever for like-minded moms to band together. Have you ever wondered why and how people branch out into smaller groups, segmenting any particular market demographic? I can't speak to the male population or any ethnic groups, but I can help you understand moms.

We've talked about two important concepts, among others, in this chapter that will allow you to understand the creation of sub-segments in today's mom market. First, the customization of motherhood being fueled by Generation X and Y mothers and secondly, the creation of peer groups based on common values. Customization will explain the "why" behind all the new distinctions of moms, and common values will explain how they come to life. I believe it's important to understand these two contributing factors because it makes the task of marketing to moms far less intimidating. Without this knowledge a marketer could easily argue that too much segmentation exists to cost-effectively tap the market. However you can argue that, although many segments of moms are popularized in the media with sexy names like "eco-moms" and "alpha moms," they are all parts of the general population of mothers who have customized their role as a mom and joined together with others who share their values. By focusing on the five core values, marketers can establish a dialogue of relevance regardless of the segment of mothers that exist in the marketplace.

Every day there seems to be a new segment of moms in the media or online. I personally think it is exciting to see motherhood evolve from the homogenous days of June Clever. Today, mothers take pride in being a slacker mom or no-drama mama and social networks have given them a platform to join with other similar women. Additionally, you can always count on a public relations specialist to give media-savvy names to those moms who don't have the creativity to do so themselves. Step aside soccer moms, let me introduce you to today's moms.

Today's Mom Sub-Segments

Mojo Moms

These moms don't believe you have to lose your pre-mom identity just because you have a toddler in tow. According to Amy Tiemann, founder of Mojo Moms, life is a lot more complex and interesting than any oversimplified idea of "Opting Out" or "Mommy Wars" that you read about so often these days. She encourages women to continue developing their own interests and goals throughout life, shaping a career path over the long run that takes family needs and personal seasons of leadership into account. One thing is certain: the world needs the full range of talents that moms have to offer! Mojo Moms seek to discover who they are and how they can continue to share their talents with others while juggling the demands of a family.

Alpha Moms

A lot has been written in the media about Alpha Moms, albeit sometimes incorrectly defining this sub-segment. The term "Alpha Mom" is often used to define mom influencers. However, as we will discuss in Chapter 4, mom influencers exist in all segments of mothers. Alpha Moms are just one of these sub-segments. "Alpha Moms" is the new term for "Super Moms" of the 90's. Think Martha Stewart on steroids. She's a well-educated over-achiever who takes multi-tasking to a new level. She is motivated by the satisfaction she earns by being in the know and the sense of accomplishment she feels by doing it better and faster than her peers. She often holds a leadership position in her community such as PTO President or Team Mom.

Beta Moms

Consider these moms the opposite of the Alpha Mom. They aren't worried about the socks on the floor and pick their battles with their children. Experiences are more important than material items and giving junior a little down time is just fine in their home. Don't misunderstand them. They are good moms and don't abuse their authoritative role but their definition of what's important is self-defined. They are the moms that might forget to sign a permission slip or return a library book but they don't fret over it.

Hybrid Moms

We've preached it for years; segmenting the Mom Market by "working" and "stay at home" categories is outdated and wrong. Two categories of "work" no longer exist and the growing number of Hybrid Moms proves it. These women, who choose to integrate a career or professional work into their in-home parenting life, are proud of their dual role. In fact, they even have their own magazine, *Hybrid Mom*, which launched with over 100,000 in circulation before the first issue was printed. Hybrid Moms put their role as mother first, yet take pride in melding a professional life into the extra hours of the day.

Moxie Moms

Fitness and fun are what brings these moms together. Their mission is to support moms in their pursuit of a community of friends, fun and fitness. The term "Moxie Mom" was coined by Susan Lavelle in 2003 and it has quickly grown to include moms with the moxie to stay fit before, during, and after pregnancy.

Ecomoms

Ecomoms are green with a capital "G"—concerned with toxins in her kitchen, organic produce and low-impact living. She is about teaching her children to value the Earth's resources and making wise environmental choices. These Ecomoms are often highly educated women who are lead by celebrity moms such as Sheryl Crow and Robin Wright Penn. These moms are not only washing in cold water and using low-energy light bulbs but they are organizing themselves in the name of their children's future. One such mom maintains anonymity but writes a popular blog for moms titled "GreenAndCleanMo" where she regularly points out healthy and safe alternatives to the bad behaviors many less-eco minded moms possess. Another mom runs "Green Mom Finds" which highlights green, sustainable products. Marketers can witness firsthand the power of word-of-mouth marketing among mothers by visiting this site. Ecomoms are organizing into groups such as Parents for Climate Protection and Ecomom Alliance, redefining the definition and reality of "mommy groups." In research by BSM Media, we found that most mothers would like to do better in the area of going green; however media has diluted for them the steps to arriving at this goal. Many moms cited a lack of true definition to marketing terms such as "organic" and "natural" to their downfall in becoming truly green. One mom told me, "The whole 'green' theme has become like fat-free was in the 90's. Everything has 'organic' or 'natural' on the label making me wonder what these words really mean to me and my family."

Yummy Mummies

Think Victoria Beckham, Kelly Ripa and Nicole Richie. These are the new pop culture driven moms who aren't ready to give up their taste for fashion and style just because they are toting diapers in their designer bag. According to London-based Polly Williams author of *Yummy Mummy* (Hyperion, 2008) a yummy mummy is "what happens when hipster 30-somethings breed." The growth of this segment has been fueled by the media focus on celebrity mothers like Angelina Jolie, Halle Barry and Jennifer Garner who all, with their little bundle of joys on their hips, retain their pre-pregnancy figure and flare. Like many other Generation X moms who are customizing their definition of motherhood, these women don't believe that childbirth or adoption has to mean losing their personal identity. Companies trying to reach this segment of mothers need to consider the style of their product as to how it fits into their self-image of parenthood. The best example of this is the diaper bag. Long gone are the Winnie the Pooh quilted bags in baby blue. Instead these yummy mummies tote Kate Spade bags, which from the outside appear to be a typical over-sized black bag with the functionality of a diaper bag hidden inside.

Momprenuers

They don't consider themselves working mothers yet they own businesses. Momprenuers are business owners who put their role as mother before their identity as entrepreneur. Their motivation to launch a business is largely fueled out of their desire to become an in-home mother, enabling them to spend more time with their children. The companies they run range from eBay auctions to franchise businesses. Two of the most famous moms among this group are Lisa Druxman, Founder of Strollers Strides and Julie Aigner Clark, Founder of Baby Einstein. Both of these moms created well known, revenue-generating brands on a global scale from the confines of their own home with kids beneath their feet.

These are just a few of the mom groups you'll find in today's market. Remember that regardless of the name they hook to their identity, these moms all possess a link to the five core values described earlier.

It's no secret that today's Mom Market is more lucrative than others and promises marketers the chance to greatly affect their bottom line. Regardless of their generation, customizable identity or lifestyle selection, moms share five core values that can be leveraged to establish a meaningful dialogue. Now that you are familiar with your target market, it's time to turn our attention to the use of new media in which to converse with your new marketing partner—mom.

Mom Mavens are the "with" you want to include in your marketing strategy. They talk, share and engage with you and your target market.

Mom Mavens: Spreading the Word of Mom

CHAPTER

I CALL THEM MOM MAVENS℠. OTHERS CALL THEM MOM INFLUENCERS. Whatever term you use to identify them, they are the go-to moms. The moms that share information with peers, direct fellow stroller-pushers to retailers and caravan minivans to destinations—all with their single recommendation. Their product knowledge and brand opinions are so sought after in some circles that they control the destiny of some marketing efforts. One may wonder why I am devoting a chapter to Mom Mavens in a book focused on new technologies and media. The answer is two-fold. First, many of the women propelling the growth of technology in the Mom Market are, in fact, Mom Mavens. They are leading the charge in blogging, podcasting and videocasting and, because of their influence, others are following. Secondly, as Mom marketing changes from marketing to moms to marketing *with* moms, it becomes increasingly important to understand Mom Mavens.

Marketers have always known the power of word of mouth marketing, however, today's new technologies are allowing marketers to use this communication channel more effectively. In fact, where "word of mom[SM]" used to be an almost unexplainable phenomenon, today marketers can manage it as a calculated marketing strategy.

Winning the heart of a Mom Maven is a marketer's dream come true. Her loyalty can earn you thousands of dollars in sales, hundreds of mentions throughout her day and perhaps millions of impressions online in blog, podcasts or videos. Nothing beats a third-party Mom Maven endorsement of your product. According to research conducted by BSM Media, 70% of moms purchase product based on another mother's recommendation. It's viral marketing at its best, but you have to be willing to relinquish control of where your marketing message is delivered. For instance, if you engage moms to speak about your product, you can't complain if she decides to blog about it on a website you would otherwise avoid with paid advertising.

A great deal is being written and spoken about word of mouth marketing. First, the Word of Mouth Marketing Association, WOMMA, formalized the marketing tactic that many of us have known to be the strongest form of marketing. Then Tremor launched Vocalpoint as their delivery vehicle for word of mouth marketing in the mom market. Now marketers are scrambling to figure out how to leverage Mom Mavens.

Word of Mom[SM]

"Word of Mom," as we like to call it, happens not only online, but off line as well. It happens in local communities with programs that engage moms with products and messages to share with their peers. Online Word of Mom provokes mothers to forward emails and give recommendations to other moms within their virtual communities. BSM Media designs programs that first identify the right mom for your product and then arm these moms to spread the word with your message. We measure their success and report the results.

In a recent Word of Mom program, 82% of moms reported telling 30 to 50 other moms about our client's product. During another program, BSM Media was able to garner 1.5 million Mom Generated Impressions for their client's brand. These are great results, but the most important element is the delivery channel—another mom. As I stated previously, the majority of moms (70%) purchase a product based on another mom's recommendation. If you need even more numbers to support the power of Word of Mom Marketing, consider the results of 2008 research released from BabyCenter, the largest online resource for pregnant women and new moms. According to their survey, their audience of mothers engages in 109 word-of-mouth conversations a week. Sixty percent of conversations among these mothers-to-be and new moms include product recommendations. Word of Mom marketing *IS* the most effective means of marketing for your product or brand.

The optimum engagement for marketers is to gain the seal of approval by Mom Mavens, but have you ever wondered what motivates a mom to be your brand evangelist? Or why other moms respect her so much? Well, you are about to find out in this chapter, as we focus on Mom Mavens and how marketers can identify, engage and empower them to sell their brand.

Maven's Motivation

Mom Mavens, as we mentioned, are the "go-to" moms within their peer group. Sitting in the bleachers at almost every little league game, you can find her. She is the mom that others turn to for the score, as well as recommendations on retailers and dinner ideas. She proudly distributes her thoughts, opinions and most of all, little-known knowledge and facts. Others clamor around her and agree as she dishes out her insights. Within the pyramid of her peers she sits on the top, pointing the way to more information and recommendations. It takes time to know all the things she knows. She must surf the Internet, read product labels, comb magazines and even watch a great deal of television. But in the end, this mom can tell you the latest lipstick color, Back to School trends and what shoes Angelina Jolie's children are wearing for the holidays.

What motivates her to always be in the know? Remember the five core values of a mom we discussed in Chapter One? Well, this mom is highly motivated by her need to share, nurturing relationships, and demonstrating that she can do it better and simpler. She enjoys her position at the top of her peer pyramid because it gives her that sense of accomplishment that she strives to obtain. This is the self-made Mom Maven. Her reign at the top is quite intentional. There is a second type of Mom Maven who earns her position by mere accident. We will discuss her as we look at the qualities that make up a Mom Maven. What you will soon see is that some moms, by virtue of their social situations, become Mavens purely by coincidence.

Making of a Maven

Mom Mavens earn their spot at the top of the pyramid of peers by their extensive breadth of knowledge. In the late 1990's, my team at BSM Media and I began to see some commonalities among Mom Mavens. More and more companies were asking us to engage influencers in marketing programs for them, which meant we had to seek out the most influential moms in the nation. We leveraged many existing relations but also considered referrals from other Mom Mavens. What we found was, first of all, Mavens can not only recommend the best stroller to purchase for a newborn, but they can also recommend other moms like themselves. Soon we had relationships with thousands of Mom Mavens across the US. You can ask me for a name of a mom influencer in San Diego and I can give you ten without blinking an eye. My first call would be to Cindy Robinson, mother of four, Board of Director for Parents Connect and business owner. In Chicago, I'd give you the name of Julie Marchese, mother of four boys including twins, church group leader, part-time teacher and team mom. The more we networked with Mom Mavens, the more we noticed common traits or qualities among them. In fact, we identified 18 shared behaviors and traits that make them an influencer.

We began plotting these qualities and eventually developed the Mom Matrix™, a system of 18 common behaviors that position a mom to be a distributor of marketing, product and brand information. They are the behaviors that make her most likely to be a true Mom Maven over a mom who simply speaks to a few other moms over the course of her day. Not every Mom Maven exhibits every trait, however, our experience and observations illustrate that some combination of these qualities ultimately propel her to the status of Maven among her peers. We will explain why after we look at some of the traits.

Among some of the characteristics included in a Mom Maven are the following:

- Mothers of multiples
- Mothers of physically challenged children
- Mothers who have suffered a tragedy
- Mothers who hold a leadership position in their community
- Moms who are published, even self published
- Moms who control a media outlet such as website, newsletter or blog
- Moms with more than three children
- Moms who run a business
- Mothers who have exhibited a strong non-profit commitment
- Moms of preschoolers

A combination of two or three of these behaviors allows you to identify moms who have strong established social networks. In word of mouth marketing, these networks are leveraged as outlets for her to share marketing messages. At BSM Media, we use our Mom Maven Matrix to screen potential Mavens and rate them against the behaviors we seek as influencers. We feel that it is not sufficient to merely ask moms if they are willing to talk to other moms about a product or to set a number of interactions she conducts with other moms as criteria. The reason for this belief is that there are moms who make it a business or habit to enter sweepstakes online, participate in focus groups or register to test products. It's so well known that even their peers know of their practices. Do you think their opinion about a health care provider means as much as the opinion of a mom with an Autistic son? Of course not. As marketer, you want to engage the most respected authority in the peer group.

Let's take a minute and examine why some of these traits produce a Mom Maven. Moms of multiples are viewed as a go-to mom because, in the eyes of her peers who parent singlets, there is a perception that it takes more know-how, a greater need for convenience and ease, and additional products to parent her twins or triplets. Thus if the mom of multiples has greater needs as a mother, the deduction by her peers is that she must know more about parenting. I remember, as the mother of three babies under two years of age, mothers of singlets gawking in amazement that I could do with three what they attempted in exhaustion with one. It's this feeling of amazement that puts the mom of multiples on the chart to becoming an influencer.

The same kind of mom-logic applies to moms of physically challenged children and moms who have survived a tragedy, whether it's a divorce or family illness. The thought by her peers is that somehow it takes more knowledge to weather their storm so they must possess some kind of insight not readily available to moms with more simplistic lives. These are the moms who accidentally become Mom Mavens because of circumstances that are often uncontrollable.

The emergence of technologies such as self-publishing, blogging, podcasting and video casting have produced a whole new breed of Mom Mavens. These self-declared, self-made experts have utilized technology to build up audiences of moms who respect and admire their opinions. They control publications, websites, e-newsletters and blog posts, and through those, the audiences who read, listen and engage with them. These women have taken the role of PTA President to the extreme. They not only influence local mothers, they can touch moms around the globe. Moms such as Kit Bennett of Amazingmoms.com are followed by millions of moms who plan birthday parties, while Mia Cronan of Mainstreetmoms.com entertains thousands of mothers across the US with helpful tips and advice. Elizabeth Pantley, author of several parenting books, travels around the world speaking to mothers about baby sleep and potty training. All of these women are mom influencers and their impact on other mothers is immeasurable. Mom bloggers make up a large portion of Mom Mavens today. I've devoted an entire chapter on these women and how companies can engage them to deliver their brand message.

Do not let it be forgotten that the good, old-fashioned Mom Mavens who use little to no technology in their community interactions still exist today. They are the Girl Scout leaders, PTO presidents, day care providers, teachers, bake sale chairperson and any others who hold leadership positions and influence their peer groups. These mothers, although possessing a smaller reach, can nonetheless impact sales and brand awareness in local markets.

Mining Mom Mavens

The most commonly asked question I receive by marketers when it comes to Mom Mavens is, "Where do we find them?" Well, of course you can pay someone to find them for you, but I caution you to learn what process they are using to determine if a mom is a Maven. There are plenty of companies out there who are paying people to sign up and build databases. You can imagine how startled I was recently when a young unsuspecting gentleman called me to solicit my services in building his company's influencer database. His proposition was to pay me $2 for each mom I convinced to register. My skin was crawling as I turned down his offer. Today, I watch from the sidelines as I see this same company sell their database of influencers to unsuspecting, yet eager marketers. My high standards of professionalism will not allow me to spread the word about the source of their "influencers" although inside I am screaming the gospel. Companies who sell mom influencers typically charge between $100-$150 per influencer.

If mining your own Mavens is more your style or more in line with your budget, there are a few tactics you can use to find mom influencers. First, identify the right Mom Maven for your company's needs and goals. Not all Mom Mavens are influencers in all circles. For instance, if you have a prenatal product, your best influencer is a mom of a toddler or a mom who is pregnant for the second or third time. Moms look up to moms who have already experienced the parenting stage they are currently experiencing. They seek information from experienced mothers who are elevated peers.

As we've discussed already, not all Mavens are created equal. Influencers can be found in each and every sub-segment of moms, from Alpha to Beta mothers. Once you've decided the right life-stage and demographic criteria for your influencer, determine if there is a common interest you would like to share with your Mom Maven. For instance, if you are a brand manager promoting a healthy cereal, you would want to seek Mom Mavens who share the passion of eating right. Not only will this shared interest deepen your relationship with her, but it makes her recommendation of your product more credible among her peer group. Lisa Druxman, Founder of Stroller Striders, would be the perfect Mom Maven for a health food product because she lives a healthy lifestyle and leads a company with 750 health-conscience mom groups across the United States.

You've identified the right types of Mom Mavens for your brand so it is now time to find your marketing partners. Offline mom influencers can be found by scouring local parenting publications. Toward the back of these publications you can find listings of mom organizations and playgroups. These listings often post the leader's name and phone number. You can also often find calendar events for mom groups, such as Mother of Twins or Moms Meet Ups. A friendly phone call to the organizer can result in, not only enlisting her help, but also gaining referrals to other local Mom Mavens. Online, a simple search by location, interest and any combination of mom or mothers can normally turn up local mom organizations. Again, an email or phone call to the leader of these groups can uncover a mom influencer. It's very important to remember that utilizing Mom Mavens is a partnership so you should be very clear about the benefits each party, you and the mom, will receive by working together. For offline groups, you might offer to sponsor an up-coming meeting or giving the Maven some type of exclusive product to share with her members.

Marketing with Mom Mavens

Now that you have identified the right Mom Mavens for your brand, how do you engage them to help you market your product? The team at BSM Media has conducted over 1,000 Mom influencer programs, enlisting over 250,000 Mavens over the past decade. Based on the client's budget, we design customized programs to fit their goals. The good news is that, unlike other marketing initiatives, money doesn't necessarily govern the results of these programs. Our experiences working with budgets ranging from $10,000 to $100,000 has allowed us to hone in on the key elements necessary in leveraging Mom Mavens.

In order to engage these influencers with your message, the first component to supply them with is some type of exclusive nugget of information that elevates them among their peers. As we discussed earlier, this nugget provides the knowledge that motivates her to share it with her friends. Access to unknown facts, product previews and special offers are all things that keep her on top of the peer pyramid.

Next it's important to also give her something to share with others. It's great to empower her with information to spread for you, but giving her something she can actually share will double your results. Many marketers simply send samples to influencers and hope they begin telling others about the product. However this approach lacks a key element of sharing—something to share with other moms. Sharing a valuable coupon or product sample with her girlfriends will position her as hero among her peers. Think about the last time someone shared an online promo code with you or a retailer's 25% coupon. You likely felt grateful or loyal to that person in some way. This is the emotion that fuels word of mouth.

The third element of your Mom Maven program, budget permitting, is something special for her child. Moms love for others to show attention to their children. Of course, if your product is focused on the woman-side of mom, this is not as important to your program. However in a program for Cartoon Network, we provided a special gift for the Maven's child and later saw the special edition toy everywhere from school grounds to eBay. Even the latter gives your program visibility. It's viral marketing in an unexpected place.

Finally, you must give moms ideas on how they might help you spread the word. In the Cartoon Network program where we were working with 10,000 Mom Mavens, we sent among other items, branded tattoos and stickers. In the package, we included a letter to the mom with ideas as to where she could distribute these items. We took into consideration solutions that moms are always seeking for activities such as scout troops, birthday parties, classroom reward boxes and bag stuffers for event favors. Our recommendation was for the mom to send them into their child's classroom as a fun reward or use them to fill birthday piñatas. Suggestions such as these again position the mom as a hero, not only among her peers, but also to her child. Can you imagine how cool it would be to have a mom that had access to early movie previews or Cartoon Network tattoos that no one else has? Very cool! Moms love to be cool in the eyes of their children.

Another example of a cohesive Mom Maven program comes from Nestle Nesquik. The chocolate milk maker was testing a new product in several Western states. They engaged Mom Mavens in trying samples of their shelf-safe chocolate milk last year. Rather than just sending samples, each mom received instructions for holding a Chocolate Milk and Cookies Mixer. Their care package also included a plush toy for their children. The overall Word of Mom program was a huge success. Eighty percent of all mothers invited other moms into their home to sample the milk at their preview party. The best news for Nestle was that on average these moms told 20-30 other moms about the shelf-stable milk. Nestle successfully became a part of the mom dialogue going on between Mom Mavens and their peers. Sales in the test market were so strong that a national roll out occurred six months before expected.

Before we conclude this discussion on key elements of Mom Mavens, it is important to mention the most important aspect to your programs—measurement. Some argue that it's nearly impossible to measure word of mouth; however, if you set this expectation with your moms as a part of the program's description, there are ways to calculate results. BSM Media utilizes several measurement tools—from online surveys, to posted event interviews, mentions from moms, and general feedback—that also include intent to purchase.

Mom Maven Activities

There are numerous types of programs marketers can launch to utilize Mom Mavens. Among them are Mom Maven Mixers®, in-home parties, influencer campaigns and advisory boards. All of these programs should include the elements described earlier. Only the execution of these components changes from program to program.

Mom Maven Mixers®

These grass-roots marketing programs are the perfect combination of viral marketing and mom influencers, and can be conducted in 1 to 100 markets. Basically, a Mom Maven Mixer is an exclusive by-invitation-only special event hosted for Mom Mavens. We've done everything from Pedicure Parties to Fashion Shows for these influencers. During the event, it is nice to have representatives, hopefully moms, from the host interacting with the invited moms. It's a great way to put a face to your brand by letting the Mavens know that there are like-minded mothers behind your product. Samples, previews and displays can all be a part of the event. Mom Mixers should be fun and provide a platform for moms to socialize. They can also be educational. Hewlett Packard hosted Mom Maven Mixers inside of Longs Drugstores in California last year. The events had a social feel to them, but also educated moms on how to do more with their digital photos. Even during leisure time, moms are still multi-taskers so this dual approach works well. Moms left the event with coupons or items to share with their peers.

In-Home Parties

Everyone loves a party and the best part of a Mom Maven In-Home Party is that you create a platform for the moms to showcase her "Mavenism." I'm not sure that is a real word, but the party lets the Maven share what she knows, and what you supply her, with peers. In-home parties work really well in the summer months when moms are looking for boredom-busting activities. We have conducted over 100,000 in-home parities with Mom Mavens. The easier you make it for the mom to hold a fun get-together, the greater the response you will have from her. The program begins when you invite Mom Mavens to serve as party hosts.

The simplest explanation of this initiative is that each Mom Maven hostess invites her peers to enjoy an in-home themed party designed around your product or brand messaging. The hostess receives a party box that contains everything she needs to conduct her party. Depending upon your budget and your goals, the supplies in the box can be increased or decreased according to the size of the party. Should you be fortunate enough to have a large budget you may consider uber-party kits large enough for scout leaders or other mom leaders to host large grass-roots parties. Although we call them in-home parties, we always tell moms that the party can be conducted during a planned playgroup at the playground or a prearranged meeting of moms. The idea is just to create an event that gets your brand in front of moms.

The party box sent to the Maven could include an instruction sheet as well as any combination of party favors, invitations, snack recipes, product samples, coupons, company literature, decorations, door prizes and a special hostess gift. In the best of budgetary times, a Maven hostess would receive a box with all these elements inside. However for the more realistic budgets, I have a few cost-cutting suggestions. There are several pieces you can house online.

Remember every component in the box adds weight, which adds postage, which in turn adds expense to your program. An online printable invitation can be created and moms can be pointed to a micro site to download recipes, activities and games to play during the party.

The hostess letter with instruction sheet is too important to put online. It's your communication with the moms and needs to be personal and engaging. It should also include the post event measurement instructions whether it's a paper survey or URL to an online polling tool.

In-home parties using Mom Maven hostesses can be very effective. A consumer product company recently hosted in-home Mom Maven Mixers to introduce a new line of products. Not only did over 12,000 moms volunteer to host parties on behalf of the company, but the mothers who qualified to become hostesses exceeded the anticipated sample distribution by 50%. An entertainment client recently sponsored Dino-Mite Mom Parties with 500 preschool mothers. Each mom received a party kit filled with party activities, recipes for snacks, entertainment, samples, coupons and other shareable promotional items. The turnkey program resulted in thousands of word of mouth mentions and hundreds of thousands of online blog impressions. In fact, 82% of moms told at least 15 other moms about the brand.

You would be surprised what moms are willing to do with you. Last year we held painting parties in seven cities across the US. The goal was to introduce a new paint with special marbleizing qualities. Moms actually invited friends into their home to re-paint their walls. The program was very successful and although it was not the intention of campaign, it's my understanding that many Cosmos were also consumed along the way!

Influencer Programs

Moms don't have to throw a party to spread the word for you. With a compelling program, moms will carry your message, product samples and coupons with them during their active days, scattering them along the way. The best influencer programs take into consideration all the components of an In-home Party—only without the party. We send Mom Mavens a product kit with ideas on sharing it with others. This is a good initiative when your campaign is based on coupon distribution. Moms are happy to leave a handful at the school office or daycare reception desk. The best part is that a mom with influence is more likely to get approval to leave your coupons at the Pediatrician office than your average big name company.

Companies can also leverage a solely online influencer program by recruiting Mom Bloggers or other mom influencers, such as webmasters and podcasters. Even non-media controlling moms can be utilized for online programs. A great tip again—instead of sending snail mail packets to each Mom Maven, marketers can establish a micro-site with e-coupons, buttons, e-cards, and prewritten emails that can be easily forwarded to other moms and audio or video files. Whatever you load on the site, just make it easy to forward and make sure it includes a call to action.

Advisory Boards and Mom Panels

Another effective way to engage Mom Mavens is to create Advisory Boards or Mom Panels. They are a great way to put an influential group of moms up against your brand and help deliver your messages. The size of the panel is optional. We have created advisory boards of moms as large at 100 for some clients, and as small as 10 for others. Again, budget and goals will dictate this. In fact, size does not limit the effectiveness of your panel. If you select a handful of moms with the right personalities and an existing audience of followers, you can obtain as much or more success than by creating a board of 100 poorly selected mothers.

Another limiting factor to your success is the level of engagement you provide these moms with—not only product development, company personnel and marketing campaigns, but also with other moms. Remember this is the world of interactive engagement. It doesn't serve the audience well if you have an advisory board of moms with whom they cannot interact. It's about sharing ideas, but not in a one-way direction. We will look later at one company whose best intentions back-fired because they forgot this important rule.

You should select your Mom Mavens much in the same way we've already described in this chapter. Common interests, shared passion and a clear outline of expectations are essential to your success. Companies can use mom advisory boards in many ways. They make great sounding boards and focus groups for new products, ad campaigns and marketing ideas. They can also be used as spokes-people and online experts.

We created a Moms Advisory Panel for one client and the moms, many of whom were authors, supplied the company content for both online and offline publications. It gave the moms the opportunity to gain visibility among their peers and exposure for their professional endeavors while the company received free content. Best of all, it was content written for moms by moms. It was a win-win situation for both sides. In justifying the expense for a program like this, you may want to spread the expense across research, content and word of mouth marketing. Formatted properly, your Moms Panel can save you expense in all three areas.

It is worth mentioning that not all mom panels have to be created from moms outside of the company. Some of the best influencers can be the moms who work to develop and promote your products every day. Bottle their passion and experience and share it with other moms that are your target as consumers. Orbitz has done this. They recently created a panel of 20 employees with children of varied ages. Among the topics these parents discuss online, in an interactive forum, are cruise tips and babysitters on vacation.

Like any other marketing program, there are right and wrong ways to execute this type of initiative and in today's viral world of moms, the latter can create as much buzz as you hope to generate through a successful program. McDonald's Mom Panel is a prime example of this.

Missed Opportunities

In May of 2006, McDonald's announced the creation of a moms panel they called Quality Correspondents. They do have a Global Mom's Panel, but the goals of the programs differ and for our purposes we will focus on the former. According to their website, "*...the Quality Correspondents represent real moms across the country. They come with different backgrounds, questions and perspectives, but with one thing in common: They are moms who care what their families eat. The Moms' Quality Correspondents, like you, want to know that they're providing quality, nutritious food to their families. The Moms will have unprecedented access to the McDonalds' system to see how McDonalds serves millions of customers' quality food every day. They will ask the kinds of questions that you yourself want answered, and will share their experiences via online journals, videos, and photos. We invite you to join this journey by following along, and interacting with the Moms by submitting your own questions.*"

And what does the Quality Correspondent program offer the average moms? Well again according McDonald's micro site, www.mcdonaldsmom.com, "*In addition, by joining the online community, you'll receive regular updates about the Correspondents' activities*." The site has profiles of the six moms selected as well as their journals of special McDonalds field trips. Also included is a Q&A section. It seems like a great program on the surface right? Wrong. The problems with McDonald's mom program started way back at the beginning with the call for entries. When the company solicited moms to apply for the six-member board, reports are that they received 10,000 to 30,000 entries. All moms were eager to help market the great brand. I must confess that as a mom who started her teen career under the golden arches, I even applied. A McDonald's loyalist who specializes in Marketing to Moms with a radio and television show, I thought I was a shoe-in for the job. And I was willing to lend my expertise and audience without compensation. Apparently at least 10,000 other moms thought the same thing.

I filled out the application and waited for a response but one never came. In fact, I learned about the winning women only because McDonald's public relations firm sent a press release to Mom Talk Radio. It was the first and only correspondence I got from the company until the first newsletter. There was no automated thank you email, no alternative way for me to become engaged, nothing at all. As a marketer, I saw a big missed opportunity for McDonalds. If 10,000 moms raise their hand to volunteer to help me market my brand to other moms, I'm not going to turn them away. I'm going to find some way, any way, for them to feel at least minimally engaged in our company.

As a mother, I felt like I had given my time to a company that didn't even do the one thing I ask of my children on a daily basis—say thank you. I apparently was not the only mother that went away disappointed. In fact, many moms circulated emails, posted blogs about the program and one mom even set up a website calling herself the "7th Mom." McDonald's gained visibility but not the kind they were seeking. Even marketing publications wrote about the flaws in the program. The missed opportunities continue even today, as does the less-than-positive mom buzz.

To McDonald's credit, it appears that the legal department is running the program. First, the selection of the ethnically diverse members was seen as a deliberate attempt to be politically correct. Second, the panel does not allow for any interactivity between the McDonald's Correspondents and moms. In a time when two-way dialogues and interactive communication are the expectations of most mothers online, McDonald's is providing "the chance to receive regular updates." I'm not sure what value this offers moms who are expecting an online community as McDonald's promises. The definition of an online community for Millennial and Generation X moms is a place where they can share and have dialogue with like-minded moms. Other than submitting questions, there is no way for outside moms to share or engage with the McDonald's Correspondents. Again, I am sure that the legal team didn't want to deal with the unknown elements of consumer-generated content and this was the compromise.

My final comment on McDonald's program is about another missed opportunity, and one that they may not even realize yet. The electronic newsletter that moms receive with updates cannot be forwarded. I've tried it several times myself. If a mom views something interesting, there is no way for her to share it with a friend. It's an easy element to change and one that can make a big difference. McDonald's is, however, making some changes and expansions in the program that could save their initial effort. Recently, the company's Moms Quality Correspondents made a special appearance on iVillage's "In the Loop" show to talk about their experiences in the volunteer program. During the show, McDonald's gave out $5 Arch cards and free samples of its Fruit and Walnut Salad to the audience. It was creative and fun and involved other moms. They also recently announced the formation of some local Correspondent programs which could go a long way toward engaging additional mothers.

Mickey Mavens

I always like to end a chapter on a positive note so we'll turn our focus to a company whose Mom Panel is going gangbusters for both moms and the sponsoring brand. It's the Walt Disney World Mom's Panel. In the spirit of full disclosure, although I am now engaged in working with Disney on off-shoot programs of the Mom's Panel, I was not part of the conception or selection process of the final mothers. The Disney Mom's Panel was created by the marketing team at Walt Disney World and headed up by Leanne Jakubowski. The goal was to create a forum for vacation-planning parents where they could get insights from moms just like themselves. The response to Disney's call for entries was overwhelming. In just a few short days, over 10,000 moms applied for the 12-member panel.

Disney experienced the same excitement among moms that McDonald's had just a few months earlier. The difference however is that Disney decided to find a way to engage the 10,000 moms who didn't make the final cut. They kept in contact with applicants and soon after announcing the names of the 12 Disney Mom Panelists, they extended an invitation to the other moms to join a newly created, yet exclusive group called The Mickey Moms Club. Incidentally, the group didn't have a name initially because the first entry into the interactive community was to ask the members to name their group. The participating moms selected Mickey Moms Club. In the true fashion of Disney, they surprised and delighted the Mickey Moms Club members with an exclusive membership kit mailed to their home. Today, members can submit questions, receive special emails and information, read profiles of other moms and interact with other members. Disney successfully turned what could have been 10,000 disgruntled moms into brand evangelists.

Meanwhile, the Walt Disney World Mom's Panel has generated hundreds of viewers questions, tips and pieces of advice.

Mom influencers are an integral part of Mom 3.0. These moms are often the link between mediums, content and other moms; delivering to mothers the benefits that Web 2.0 promises to bring the Internet. Perhaps they are the missing link that developers are now trying to replicate in order to take the Internet to the next level. It's not surprising. Moms know what other mothers are seeking in content, as well as the delivery channels that make it easy for moms to obtain what they need.

Mom Mavens are the "with" you want to include in your marketing strategy. They talk, share and engage with you and your target market. Ensure the success of your Maven efforts by finding the right mom influencers for your brand, engage them in a win-win relationship and give them the platform and tools to share with other mothers. Entertain, educate and empower them and you will enlist the services of your most effective brand evangelists in the Mom Market.

Moms of all types are leveraging social mediums to do more in a more efficient and fulfilling manner.

Social Networking With Mothers

CHAPTER

EVERY AUTHOR THINKS AHEAD IN THEIR WRITING AS TO HOW THEY WILL introduce a new topic in their book. I am no different. In fact, I spend a lot of time thinking about marketing with moms while running. Today the focus of my morning training run was social networking. I thought about the most important aspects of this powerful marketing tool and the key messages I wanted to deliver to you, my reader. I debated topic after topic as I pounded the black tar road in Welaka, Florida (pop. 401) where I retreat to my fishing camp as often as possible. There is only one streetlight in Welaka, but five different wood framed churches, the Log Cabin Bar and a pharmacy. It was on the sign of a small white church that I found my inspiration. The sign read: "Communities are not organized, they are exercised." What a perfect piece of advice for marketers seeking to understand social networks! I am sure that the church volunteer who hung the letters with great care never expected to inspire a marketing

A Need to Socialize

Although I contend that a mother's tendency to network is innate, it is interesting to examine her motivations for exhibiting the behavior. Social networking is such a powerful tool for moms because it hits on several of her core values and motivators. It creates a platform for nurturing relationships in a simpler and easier way. Multi-tasking, time-starved moms love platforms that bridge together several of her internal objectives.

Let's look at her need for sharing first. As we have already discussed, moms share for a myriad of reasons, from nurturing relationships to fulfilling obligations. Networking provides an easy and natural channel for sharing. Very few other segments of people come in contact with so many people during the course of a normal day with so many varied interests. Think about a mom's day, whether she goes to a traditional office or spends much of her time in the home. First, she interacts with moms and teachers while shuffling her children in the classroom. Then she may interact with service providers or retailers as she runs errands. Later, she converses with professional colleagues or clients via email, conference calls or face-to-face contact. The afternoon brings her in contact with coaches, parents and other children. Dinner time arrives and the number of interactions increases from food providers to fellow team parents, family and friends. Finally, as the day winds down she is back online interacting within social networks, exchanging emails with virtual friends and finalizing business transactions on AIM. A full day of literally hundreds of interactions with peers that she can influence with product recommendations, service provider information or parenting advice.

Marketers have realized that word of mom is the most effective way to market a product or service to mothers. Today, due to technologies such as social networks, word of mouth travels faster and further through channels that marketers can actually watch. Would you have believed three years ago that you would actually be able to see—rather than hear about—viral marketing spread, and even measure its reach more accurately? Social networking brings these marketers' dreams to life.

The growth of social networking is undeniable. MySpace.com attracted more than 114 million global visitors aged 15 and older in June 2007, which was a 72% increase year to date.[1] Facebook.com experienced a growth rate of 270% when they reached 52.2 million visitors in the same time period. Several other sites such as Bebo.com and Tagged.com enjoy the same rapid market penetration. Surprisingly, it's not just students visiting these sites. According to Facebook representatives, more than 50% of their visitors are no longer students. In fact according to eMarketer, 13% of MySpace visitors are moms and they aren't just there to spy on their children. Chris DeWoulfe, CEO of MySpace, actually declares the number to be closer to 40% of all American mothers.[2]

In an interview with Maria Bartiromo in *BusinessWeek*, DeWoulfe argued that the Internet generation of women, who were 25 in 1995 when socializing online became mainstream, are now 38 and in the midst of childbearing years. He also agrees that moms are online socializing like everyone else. I think 40% may be high for moms populating MySpace but I do know that research report supports the fact that moms are online in social communities. According to a recent BabyCenter report, 70% of moms participate in online communities spending approximately four hours a week engaged in social media.[3] Many have their own profiles. I have to confess that my first profile on MySpace was set up with the primary goal of

1 "Major Social Networking Sites Expand Their Global Visitor Base", *comScore*, www.marketingcarts.com (accessed December 6th, 2007).

2 Maria Bartiromo, "FaceTime," *BusinessWeek*, (June 2nd, 2008.)

3 dBusinessNews NewYork, article, www.dbusinessnews.com, (accessed November 27th, 2007).

monitoring my children's activity on the site. My strategy was that my children would not want to play in the same playground with their mother. Once my own children became my friends on my profile, I realized that my strategy had failed. However, I did discover a useful marketing tool along the way. Now I maintain profiles on almost a dozen social networking sites, all for the purpose of connecting and marketing myself as a brand.

According to comScore MediaMetrix October 2007 data, CafeMom.com is the most trafficked site for women in the social networking landscape. The ranking is based on page views and average minutes per visit. Other popular social networking sites specifically targeting parents include Babycenter.com, Parents.com, and Family.com. The average time visitors spend on these sites ranges from 4 to 6 minutes. The average time online increases for non-parent targeted women's sites such as Oprah.com and Marthastewart.com, with the average amount of time spent on the site exceeding 7 minutes.[4] One could deduce that women without the responsibility of children have more disposable time to be online, however, the hypothesis may be flawed since overall moms spend more time online than childless females. I believe that it has to do more with the content these sites present to the reader. Research shows that moms are increasingly replacing the act of leafing through a magazine before bed with surfing the Internet. Sites such as Oprah.com and Marthastewart.com allow mothers to escape into their womanhood similar to hard-copy publications like *Real Simple* and *Cosmo*. It would be interesting to see when the traffic to these sites spikes. My guess is that it peaks in the hours after the child's bedtime when moms begin to relax.

Community sites attract 70% of moms online. According to BSM Media, almost 60% of all moms have engaged in some type of online

---

4 Ibid

community site. Sixty-five percent of them have posted a message and 40% have read a blog on the site. Companies are becoming increasingly aware of their need to be included in the dialogue with socializing moms. eMarketer has increased its projection for US ad spending on social networks to $900 million. By 2011, they forecast ad spending to reach $2.5 billion.[5] Currently, 72% of all US ad revenues are dedicated to MySpace and Facebook. However, there are numerous untapped opportunities in the mom social network space.

Younger people use social networks such as MySpace and Facebook to explore their identities. A mom will use a social network to validate her interests, build relationships or promote part of her identity such as a book, business or cause. However, the use of social networks goes beyond connecting with like-minded mothers. Once again, mothers are pushing the utility and functionality of technology. Today's momprenuers and mom business owners use social networks to build their businesses, attract new clients and showcase their products.

One of the most popular tools for business-minded mothers is Linkedin.com. I assume as a reader of this book, you are very familiar with Linkedin and are perhaps using it yourself. However, what you might not realize is that thousands of moms utilize the site to find manufacturers for product ideas, as well as potential business partners.

Moms are also using tools such as Gmail and Yahoo! which allow her to multi-task during her online experience. Each tool provides excellent chat functionality, which moms are expanding upon to build group chats with family and friends. They are staying connected in real time with AIM and sharing photos, videos and content as well.

5 eMarketer article, "The Crowded World of Social Networks," www.emarketer.com (accessed December 6th, 2007)

Social Media Leveragers

Whenever I speak about social networking and moms, I inevitably get the same question. "Why are so many moms engaged in social networks beyond the sharing aspect?" That question is then followed up with, "Where are these busy moms finding time to do all this online interacting?" The assumption to the latter question is always that they must be stay-at-home moms who don't work. So before I go any further, let me assure you that this could not be further from the truth. Many of them are running businesses, managing households and juggling volunteer work. Moms of all types are leveraging social mediums to do more in a more efficient and fulfilling manner. In fact, BSM Media research shows that community sites attract 70% of moms online. The reason for their engagement—social networking is centered around affirmation and validation. Now let's go back and address the "why."

Remember the five motivators we discussed earlier and how important the feeling of accomplishment and connection is to a mother? Social networking allows moms, who might otherwise feel challenged, alone, frustrated and alienated, to gain validation or confirmation of everyday mom situations. Imagine you are a mom in Madison, Indiana and your big outing each day is a trip to Wal-Mart or to the park with your toddler. The scope of your relationships, personal growth and interactions are limited to a two-year-old, the Wal-Mart greeter and a few high school friends you grew up with that now share children of the same age. Suddenly by logging into www.themotherhood.com, you have thousands of friends who can assure you that it's okay to feel alone or congratulate you on your baby's first words. Or you can visit www.newbaby.com, record video of your toddler and share it with family all over the world. This leads us to the time involved in maintaining these relationships.

Some would argue that it's easier to pick up the phone, chat a few minutes with a friend or share a few minutes at the park than to log in and maintain relationships with hundreds of online friends. To the contrary, many social platforms allow moms to converse with larger groups all at the same time and, even better, at a time late at night after the family has gone to bed. Social media sites allow moms to manage their time and relationships more efficiently. However, when you ask moms, particularly younger mothers, about the amount of time they spend online, some will answer in upwards of 6 to 8 hours a day. So again, you ask where are moms getting this time?

What I've observed is a reallocation of time with a greater distribution toward social networking. For instance, a mother shopping for a product may now go to her Yahoo! group and ask for recommendations, rather than browsing multiple retail locations. Instead of watching television in the evening, she may elect to send "tweets" to her followers on Twitter.com, which we will discuss in more detail in a later chapter. Instead of attending a swim team meeting, she may swap photos with other moms on special member boards. It's a symbolic relationship between time and social media for moms. Socializing online makes her feel like her time is being used more effectively and productively and, through social communities, she is able to create more time in her day by receiving time-saving tips from virtual girlfriends. As a marketer, it's your job to have her devote a small piece of that time to you and your product. We will get to the tactics as soon as we learn more about the most popular mom-oriented social networks—straight from the moms who are running many of them.

Connectingmoms.com is a popular social network for moms. According to their own description, "ConnectingMoms.com is a social networking site for moms of all ages and stages of motherhood to connect; a community open 24/7 where the voices of embracing, amiable, and encouraging women carry you throughout the day. A means to discuss, shop and save, review products, blog, share videos & photos, and interview extraordinary moms. It's a place to find out what works for them, a place to discover what works for you. Whether you're seeking advice, looking to share your knowledge, or lend a helping hand, ConnectingMoms is the location for mothers to meet and greet, post and boast, and most importantly—make a difference." You will see as we examine these sites that common themes, such as sharing and connecting, are part of the networks goals.

Momspace.com is a more locally driven social network. In fact, it touts itself as being the only hyper-local website for busy moms. It's the goal of Momspace.com to save moms time, offer the opportunity to connect with local mothers and make it easy to gather information that is relevant to their local offline community. As with many of these types of sites, Momspace.com was co-founded by mothers, Joani Reisen and Erica Rubach. They say that the mother of invention is necessity and we see this over and over again online, as moms create the tools they need to manage motherhood.

Mommybuzz.com is run by Moxie Moms, who we met earlier. It defines itself as a place where busy moms can go to stay connected, stay in touch with friends and get involved with mom groups that share their interests. The website allows mothers to do this by offering real-time chats, journaling, blogging, photo storage and classified ads.

On a larger scale, MothersClick.com is a more populated social network. What is unique about this site is that they are also an online service provider developed exclusively for active moms and their groups. They provide a free online tool, designed to be a unique parenting resource, to help moms to promote group-building, knowledge-sharing, and community. They've built an infrastructure that can be used by new and existing groups of moms, clubs and organizations. They even tout the time efficiency of using social networking tools online and offline. It's a return to the symbiotic relationship analogy of time and social media for moms that we discussed earlier.

I mentioned Cafemom.com earlier as the largest and fastest growing social networking site for moms. According to company reports, they received 99 million page views and 2.1 million unique visits in October 2007. Within their community, moms have created over 15,000 groups around the topics of recipes, photos, marriage and relationships. The site allows mom to create fully customizable profile pages, friend networks and journals.

Although many social networks are focused on bringing mothers together online, there are some that join moms offline. iPlayGroups.com and Meetup.com are two of the most well known. iPlayGroups.com was founded by Allyson Phillips and her husband Shandley. Moms are able to connect with other mothers in their area to organize offline gatherings in their local communities. iPlayGroups.com has over 8,015 members registered on the site, with over 800 playgroups ranging from general to special needs children, with even some specific groups for working parents. Mothers are only part of the Meetup.com community, and there are literally thousands of Mom Meetup groups. They are created by geography, lifestyle choices, life stages and parenting styles. Among some of the most common groupings are Moms of Twins, Single Mothers and Military Moms.

Both of these sites are great resources for identifying local mom organizations or influencers. Simply Google your preferred city or state and you most certainly find a Meetup group. You should be warned, however, that some of these groups are private and do not welcome outsiders; they are well identified so that you can respect their privacy. For the more open groups, I've found them receptive to an introduction email, and also willing to sample products with their members.

Just as I encourage marketers to engage in a dialogue with moms, I want to turn to some of the most experienced mom social network creators for their thoughts on the subject. It's always important to listen to moms and here I'll lead by example.

The Motherhood.com

Founders, Emily McKhann and Cooper Munroe

Q. What were your motivations behind launching the site?

A. In 2007, after turning our blog (Been There, www.beenthere.typepad.com) into a clearinghouse for in-kind donations following Hurricane Katrina (The Been There Clearinghouse, www.beenthereclearinghouse.com—named the most inspirational blog of the year in 2005) we learned firsthand the power of mothers using technology to find solutions. We built TheMotherhood.com as a way for mothers to find each other to help make a difference in their own lives, for each other and for their communities.

Q. Why do you think there has been a move from content sites to social media sites?

A. Since time began, mothers have shared information with each other about everything—getting a baby to sleep, dealing with a teenager, balancing work and home—or have just provided a shoulder for a friend who needs it. The power in finding solutions and making progress in your own life depends on a two-way conversation, not a one-way, static piece of content. The Internet was made for mothers—we tell stories, we give support, but we don't necessarily have the time in our busy schedules to do all the things we want to do. The Internet is ready for us when we are.

Q. If moms still want information, tips and advice, is the content model now extinct?

A. Moms trust tips and advice from other, real human beings. Mothers are savvy enough to know the difference between processed, canned information and real content that comes from someone who knows what they are talking about and has "been there." That goes for companies too. Mothers would much rather hear from a corporation that can speak directly to them, person to person, from experience. The moms we know don't pay one bit of attention to processed, fake-sounding marketing messages that have been user tested, focused grouped and edited through five miles of red tape in order to get to a consumer's ears.

Q. How can a company use a social network to connect with moms?

A. If a company can have a conversation with mothers that gets as close as possible to an experience like sitting on a sofa talking to each other person to person (with a genuine authenticity and caring on the company's part), that is a really valuable start to creating a solid relationship. The right kind of social network can come very, very close to providing that kind of opportunity.

Q. What opportunities does your site specifically offer to marketers? Can you give some examples of things you've done?

A. We are rebuilding The Motherhood.com as we write this (even though we were nominated for a Webby this year, the site, for us, does not nearly do what we need it to—our re-launch is expected in Fall 2008) and the layers we are adding are wholly unique innovations aimed at enhancing the connections we can establish with each other, and providing moms the offline and online tools she needs to make her life a little bit better.

Q. Do you see moms being a part of more than one social network?

A. For every type of person, there's a social network. Ravelry.com is a popular social network for knitters. Club Mom certainly has a big user base and we see many mom-bloggers on Facebook and

Twitter. What remains to be seen is the social network that captures the imagination of moms who aren't using social networks or reading blogs—that's who we want to see on The Motherhood.com as well.

Q. Where do you think social networks will be in the next five years?

A. If the projections are correct, most moms will use social networks in some way in five years. Obviously, the young women who grew up on MySpace and Facebook have fully integrated social networks into their lives, and will continue to do so as they become mothers.

Q. Do you think moms want companies to be a part of their social networks?

A. If a company wants to establish an authentic relationship with mothers, yes. Moms appreciate when a car company donates to autism research, or a consumer product company genuinely wants to hear from moms how to make their products better (and they deliver on the improvements), or a corporation sponsors something moms love, just because they want to support moms. Women notice and are grateful for the efforts.

Q. Is there one mistake a company can make when playing in the social media arena?

A. Not listening, being too "corporate" or clearly just trying to sell something and not interested in being a caring, interested part of mothers' lives. Customers are no longer just the customers; they are defining the brands for themselves. If a company is not a part of the conversation, the customers will define a brand in any way they want.

The Mom Salon.com
Founder, Jennifer James

Q. Why did you create TheMomSalon.com?

A. I originally created TheMomSalon.com in 2005 when I realized that mom blogs were going to be a huge hit in the years to come. As it turns out, I was a little prophetic—although there were many others who were as well.

In 2005, there were no web directories for mom bloggers. I wanted to create the first place where moms could submit their blog information and find other mom bloggers, based on categories ranging from academic mom blogs to writers. So, if a mom wanted to find a pregnancy blog, she could do that. Or, if a mom wanted to find other mom bloggers who are photographers, for example, she could do that through TheMomSalon.com, which is a category-based directory of over 1000 mom blogs.

I also created the first (*at least I think it's the first*) social network specifically for mom bloggers that isn't a part of a larger brand: the Mom Bloggers Club, www.mombloggersclub.com. To become a member a mom must have a blog, even if it's a day old. The Mom Bloggers Club is an active online space where mom bloggers give advice about blogging, share tricks they've learned about how to drive more traffic to their blogs, talk about the ins and outs of accepting ads, and simply make friends and find new blogs to read.

Q. What's the focus of a social network for mom bloggers? Do you think moms belong to more than one social network?

A. The primary focus of the Mom Bloggers Club is to offer moms one place where they can connect with other mom bloggers to get tech tips, blog advice, promote their blog, drive traffic to their blogs, find new blogs to read, and make new friends. I wanted to create a niche social network where moms can get down to the nuts and bolts of blogging and social networking and little else.

Without question, moms belong to several social networks. There are so many that it's almost a necessity to belong to more than one. Some, like Twitter and Plurk, offer better conversational capabilities than others and that's why they are so popular.

Q. Do you believe moms want companies to be a part of their social networks?

A. I think moms don't care one way or the other unless companies try to be sneaky about courting them by either setting up member profiles and then "friending" them or following moms in Twitter and Twittering them to death about sales and promos.

Q. What is the best way for a company to get involved in marketing through social networks?

A. It's my belief that companies who want to market through social networks need to be very creative. The nature of social networks is so conversational that it's very easy for users to glance over traditional banner ads and never really see them. That's why it is important for companies to get a real handle on the users in a given social network and figure out what will capture their attention. They need to elicit and create conversations instead of simply opting for brand awareness.

Q. What opportunities do you specially offer to marketers on your sites?

A. I offer banner advertising and partnerships (co-branded and sponsorship) to marketers. And in all honesty, because Web 2.0/3.0 is so new and forever changing if a marketer comes to me a unique idea, I'll definitely consider it.

Who's Seizing Online Marketing

As the popularity of social networks has grown within the Mom Market, a few companies have stepped up online to seize the opportunity in areas such as research and marketing. These companies have created social networks for the sole purpose of selling the opinions of the moms who frequent their communities. I have to admit that I've been skeptical of these companies from the very beginning. My fear for them has always been that the authenticity of what they are doing may one day come back to haunt them in a negative way. I also fear that they will be unable to maintain an active community of moms without the benefit of a mom face behind their technology. I use Tremor's Vocalpoint as an example of this.

Tremor, a word-of-mouth company owned by Proctor & Gamble, created their community of moms after finding success in the teen market. Vocalpoint, the mom social site, offers moms product samples, the opportunity to discuss topics ranging from entertainment to products and the chance to influence others through a series of online surveys and polls. Tremor is a revenue-driven business that sells access to these moms and their opinions to other companies. It's never disclosed to the moms who join, myself one of them, that their insights are being sold to other for-profit companies. Tremor makes money on my opinions and I get a free sample once a quarter. I have always wondered what the fallout is going to be when their mom members realize this business model. Authenticity is so important to moms that it could spark some hot debate and backlash in the blogosphere. A community has to add value to the life a mom.

Going back to the church sign in Welaka, companies who want to organize an online community for moms must have a plan to exercise that community and allow it to grow organically. One of the best corporate executions of social networking I've seen is "In the

Motherhood," www.inthemotherhood.com, produced by Suave and Sprint. It's no surprise that the site achieved 5.5 million views in its first year. In the Motherhood creates a unique platform that allows moms to share their stories with millions of other women. The shows are great and change regularly. The online mini series stars Leah Remini, Chelsea Handler and Jenny McCarthy and presents humorous story lines that are universally interesting to moms. The site keeps moms connected with each other and fits perfectly with a brand like Sprint, whose products help moms stay in touch with her family and friends. In the Motherhood allows mothers to connect directly, sharing tips and strategies on managing their busy lives. It also engages moms further with programming, allowing them to suggest storylines by submitting real life stories in simple paragraph form. The online community then votes on its favorite submissions and the winning entries are turned into webisodes. The website also has a gaming area and online community where moms can exchange recipes, activity ideas and opinions in general. Of course, the homepage also has links to special promotions for Sprint phones and Suave products.

I like the fact that moms can customize how they interact on In the Motherhood. They can write, listen, vote or watch. This is extremely appealing to Gen X and Gen Y moms, who particularly want the ability to customize their relationship with a brand. They want to be in control of their interaction with your company.

Leveraging Social Networks

Now that we know millions of moms are online meeting up and engaging in social networking, let's turn to how you become part of the dialogue. The opportunities are as numerous as the moms you have the potential to engage with your brand. They range from sponsorships, contests, site enhancements such as content and technology, and good old-fashioned partnerships.

I recently listened to a presentation by Jon Stross, who is responsible for international growth of BabyCenter. He described a promotion with Garanimals, the clothing retailer. The goal of his client was to bring the brand into the conversation with moms of toddlers and preschoolers. The natural tendency would be to throw up some banner ads on pages serving up the target mom, but BabyCenter understood the importance of integrating the advertiser into the dialogue with a multi-prong approach. Ultimately they designed a discussion group around the topic of "Is 3 the new 5?" and engaged moms in an active debate online. The themed pages were sponsored by Garanimals and were followed up with a sweepstakes that generated a large database made up of their target market. Even better for the advertisers, the moms opted in to receive on-going information from Garanimals because a meaningful relationship had already been established.

CafeMom, that we discussed earlier, supports moms by offering a place to get advice, support, or just relax. They tout marketing opportunities that help advertising partners appear relevant to moms. Marketers can sponsor custom widgets, profile pages or develop custom groups around themed topics. A visit to Cafemom.com will allow you to see that many of America's top brands are investing in banner ads on journal and mom group pages. Although I am sure that they are producing millions of online impressions and are easy for media buyers to purchase, I would argue that more effective marketing initiatives would engage moms in a two-way dialogue with their brand. Sponsoring a discussion forum or posting product updates in a chatterbox would be far more effective in becoming a company who relates to Mom 3.0.

Gabrielle Blair, co-founder of Kirtsy.com, a social network we will discuss in more detail later, believes marketers need to take their promotions and campaigns to a higher level of authenticity. According to Blair, the best marketing opportunities happen when marketers work with the social network founder to identify common goals, such as traffic building and word of mouth buzz. She suggests, rather than just giving away a product to one mom, why not engage the entire community in helping develop new product lines or campaigns? Allow moms to give input on taste by sampling or helping to develop a product concept.

So I know what you are thinking right about now, "There is no way our legal department would let any of this fly." And you are probably right; they won't let it fly if you present it without a firm case for testing the waters of social media. Your list of objections to overcome will most likely range from, "We are a traditional brand with traditional methods of marketing," to "Customers submitting ideas would put us at risk of liability." I assure you as someone who works with the biggest and most long-standing brands from Johnson & Johnson to Disney, that if there is a will, there is a way. It's not impossible to convince management that your marketing programs are working when the bottom line numbers are higher, is it? Like in any approval process you sometimes have to take small steps, showing numbers that support your case along the way. The marketing managers I've seen successfully convince their legal departments to allow them to test social media programs have presented mini programs with very clear benchmarks and limited customer engagement.

Engaging Employees

Marketers often use trusted employees to project the voice of their brand. One example of this is Precious Moments, the collectible company. It's a very traditional, well-established brand that achieved success by producing popular themed figurines and collectibles that their customers fell in love with. Like most companies, they produced a product and used traditional media to push it out to the customer. Several years ago, they realized that their brand loyalists were an aging population and that it was necessary to introduce Precious Moments to a younger consumer. In other words, it couldn't survive as a product that most Boomers remembered from their grandparents.

Imagine the paradigm shift that needed to occur to target today's Gen X mother and her children. Precious Moments' plan was to enter the world of social networking and virtual gaming. In 2007 they engaged BSM Media to create and launch PreciousMoms.com. It was to be more than a mom's portal offering Christian-minded mothers a place to exchange ideas; it was a conversation starter for their second new endeavor, The Precious Girls Club, www.preciousgirlsclub.com. The latter site is a virtual world for the tween daughters of the Gen X mother they are speaking to on PreciousMoms.com. As they were developing the virtual world gaming site, Precious Moments used the mom site to provide day-to-day development news. Additionally, they engaged a female of their executive team to blog about new ideas, storyline concepts and the introduction of new characters. They also used an online survey tool to allow moms to vote on the names of characters appearing in The Precious Girls Club book series and online destination.

The strategy was to engage moms in the new brand before it even went live. For almost a year, they kept up a dialogue with content until the entire company, even the legal team, was comfortable in sharing with consumers. They measured site traffic, incremental sales and customer engagement in comments, surveys and feedback. They were also able to show a savings in research costs because they were going straight to the consumer with online polls rather than paying a traditional research company. Best of all they built a database of moms who were ready to engage their daughters in The Precious Girls Club when it launched in September 2008. It's worth mentioning as well that the content that PreciousMoms.com uses to fill its pages is 100% consumer generated through partnerships with mom bloggers who share their brand values. The entire site is built on a winning strategy for the company and a meaningful dialogue with the consumer.

Let the Moms Decide

If the steps Precious Moments took still seem too big for you, consider Disney. This is a case of a huge, global, conservative brand who took a tiny step into engaging with moms in social networking and it paid off big. I have to attribute the brilliant idea to Duncan Wardle, Vice President, Global Public Relations Integration and Walt Disney World Public Relations for Disney Parks. We were sitting around the table trying to determine the name of Disney's newest social network of moms when Wardle spoke up and suggested that we let the moms decide. You can imagine that the responses included everything from trademark issues to available URLs and that each statement started with "what if." Persistence prevailed and Disney offered up several choices of names to their moms with the legal department's blessing.

My point here is that entering the social networking arena doesn't have to be an all-or-nothing proposition. Start by examining your company's tolerance to sharing and then manage the information flow with internal resources. Finally, use an old mother's trick and give your consumers enough choices to select from, but make sure they are choices you can live with them selecting to share. The most important thing to remember, and you will hear this again and again in the next chapter on blogging, you must execute your programs in the most honest and transparent manner possible. No matter how small or large your social marketing campaign, if you are socializing with moms, they want open and honest relationships.

As moms continue to be stretched in dozens of directions each day, I believe social networking will grow in importance to her. These social networks will allow her to connect more effectively with her friends and colleagues and as her needs change, she can move within these online communities to find new mothers with similar life stages or phases. Your challenge as a marketer will be establishing your brand as an active member of these communities. It may be by engaging mothers to comment on product ideas or express their opinions through online polls. It might be as bold as enlisting the help of employees to blog or post about company progress and ideas or as small as asking moms to rate an ad campaign.

Whatever your strategy, it's important to remember the words of the little church in Welaka, Florida and exercise your relationship with moms. Keep the conversation moving, rather than organizing a plan that sits idle while moms share their thoughts about your competitor's product. Social networks give you a place in a mom's playground, and once you've found your way around the sandbox, it could result in the most lucrative conservation you've ever had with a consumer.

Blogging leverages a natural behavior of moms that can predict the longevity of a marketing strategy.

Mining the Blogosphere

CHAPTER

IN OVER A DECADE OF FORMALLY STUDYING MOMS, I HAVE NEVER SEEN as much of a frenzy surrounding one type of marketing initiative as I've witnessed recently with mom bloggers. In the recent history of marketing to moms, there have been certain initiatives that gather attention of media, marketers and the consumer. Few will forget the popularity of affiliate marketing in the heyday of the Internet. It was a popular way for companies to form an alliance with mom webmasters that included a revenue stream for the mom and access to the moms' peers. Freebie offers and online coupons soon followed giving moms access to special discounts they could electronically share with their friends. Chat rooms and message boards created forums for dialogues between moms, brands and experts. Technology evolved and then podcasting became a popular way to connect with moms. Attention then turned to consumer generated content as word of mouth marketing formalized itself into a national association and *Time* put YOU on the cover in 2006 as Person of the Year.

Mommy Bloggers

Through all these stages of marketing to moms, however, the media has never focused so much on the involvement and engagement of companies and moms as it has with the phenomenon of mommy bloggers. Companies have never scrambled so quickly to understand mom bloggers. I've watched media buyers, public relations firms and advertising managers blindly spend unallocated budgets on mommy blogger programs with little regard for reach, impressions or other measurements. Adding to the rush to play are the accidental business women formerly known as mom bloggers. Mothers who simply started a blog to chronicle their journey as a mother or to find like-minded women with children are suddenly receiving hundreds of product samples, special event invitations and free merchandise in their in-boxes.

In fact, the present activity in the mom blogosphere makes this chapter the most exciting to write, yet the most difficult. My fear is that the mom blogosphere is changing and evolving so quickly that what I write today will not be relevant six months or a year from now. However, I remind myself that there is a great deal of knowledge in understanding the evolution of a medium, as well as knowing how to execute successful campaigns with it. As I write this book, blogging is the most explosive marketing initiative in the marketplace today; however I feel it is only halfway to maturity. I believe the progression of mom blogging as it approaches maturity will affect, not only how marketers of the future will deliver their messages, but also what consumers expect from companies when they deliver that message to their peers. The days of free word of mouth marketing might be coming to an end. Only time will tell.

Today, according to BlogHer research, more than 36.2 million women participate in the blogosphere weekly. It is estimated that about half are women with children. Twenty-one million of them publish or read comments at least once a week. The majority of mom bloggers are Generation X mothers who post about their experiences as moms. More than 40% of women consider blogs a reliable source of advice and information and over 50% admit that blogs influence their buying decisions.[1] It is unknown exactly how many mommy bloggers inhabit the blogosphere, but estimates range from 100,000 to half a million. Technorati, an Internet site that tracks blogs and social media, estimates over 6,000 mom blogs and growing which is actually only a fraction of the estimated 12 million Americans who maintain blogs. Six percent of the entire US population has created a blog. According to Quirk's Marketing Research Review, 57 million people read blogs with more than 1.4 million new blog posts every day.[2] The challenge in calculating such a number is that many blogs are abandoned and inactive when the writer loses interest or lacks the time to update it. Generation Y women, interestingly, are more likely to abandon a blog, but are also more likely to start more than one blog at one time.[3] According to Blogads, a blog ad distributor, the average visitor to a mom blog is 29 years old with an annual income of $70,000.

Generation X mothers are most active in blogging. Their motivation, and surplus of topics related to parenthood most likely contributes to their loyalty to their blogs. Most mom bloggers have only been in the blogosphere for less than twenty-four months.

There are several networks of mom bloggers. BlogHer is a community for, and guide to, blogs written by women. The BlogHer network has 1100 blogs–and growing. They have recently created an ad network by gathering their blogs together and offering them to advertisers. They are so formalized that they have an annual conference that attracts thousands of women each year from across the US.

---

1 Wright, Susan. "BlogHer Statistics," *US National Census Data Projections*, June 2007, www.compasspartners-llc.com.

2 Davis, Hugh and Mike Oberholtzer. Quirk's Marketing Research Review, "

3. Wright, Susan. "BlogHer Statistics," US National Census Data Projections, June 2007, www.compasspartners-llc.com.

Similar networks of blogs are being created in the mom space. Jill Asher and Tekla Nee, some of the founders of Silicon Valley Moms, have replicated the success they have seen with their group of 40+ mom bloggers in Northern California. Today, they have mom blog networks in Chicago, New York, Washington DC and Atlanta with more cities planned in the future. Each local mom blog network has at least 30 mothers who maintain blogs. Some write with a local flair while others focus on global aspects of motherhood. The Mom Salon and Parents Blog Network also offer a community of mom bloggers to companies looking to reach a large number of bloggers in one stop. Another site worthy of mention is Kirtsy.com, an aggregator of blog content which focuses on female writers.

Why Blog?

There are many reasons that moms blog. It seems funny to refer to the early days of a medium that is so young; however in the early days of mom blogging, many moms began journaling online out of a desire to chronicle their journey through motherhood and share it with other moms. The question I am asked most often by marketers trying to understand what motivates bloggers is, "Why would someone want to keep an open diary on the Internet?" However, the answer can be found in the question.

Marketers don't understand why moms would want to put themselves out there to other moms, but in reality it's the motivation. Moms are looking for ways to connect with other moms. As busy as their daily lives are, motherhood can be a very lonely experience—particularly if your world largely circles around your child's needs. In focus groups, moms describe feelings of alienation, loneliness and uncertainty. Asha Dornfest, creator of Parent Hacks says she's motivated to blog because of the conversations she can have with moms all over the world.

"Parenting can be lonely work, especially when times are tough.

Parent Hacks reminds me that we're all in this together," says Dornfest, who started her blog in 2005. Today she has over 50,000 visitors per month with 20,000 subscribers via email and RSS feeds.

The appeal of blogging is the perfect tool for the Generation Y and Generation X mothers. Its functionality aligns very closely with the character traits of these two cohorts. First, blogging provides immediate gratification. Women of these generations were raised on instant gratification and expect it in almost every area of their life. This stands true emotionally as well. A mom expresses her feelings on her blog and within minutes she can have thousands of other moms either validating her emotions or telling her to get over it. There's no waiting and wondering what others will think. Response is almost immediate; particularly with the emergence of micro-blogging which we will discuss later in this chapter.

The second reason that blogging fits nicely into the personality of Generation Y and Generation X mothers is its ability to be used to customize motherhood. As we discussed earlier, these women are holding on to more of their personal identities than their Boomer predecessors. Blogging allows moms to connect with moms who have similarities—from their location to their passions, education, hobbies, beliefs and values. If you are a Christian mom, you might follow Lori Seaborg's blog, Just Pure Lovely, www.justpurelovely.typepad.com, and if you are a New York mom you may follow the posts on Mommypoppins.com where Anna Fader celebrates her New York upbringing by sharing the fun of raising a baby in Manhattan.

Third, blogging gives moms a platform for cause marketing and philanthropic support. This is particularly important for the Millennial generation of mothers. Educators and social institutions drove this group to make a difference in their younger years, and as adults many still support the same causes. Many of these moms use their blogs to connect with moms who share the same priorities or values. During the 2008 election, many moms with Democratic views could trade thoughts on Momocrats.com. Even famous mothers such as Elizabeth Edwards could be found blogging on this site. Beyond political values, there are blogs that support Going Green, Global Warming and Water conservation. All of these blogs create a platform for moms to share their cause related views.

Finally, blogging as a tool allows moms to retain more of their own individualism. Yes, many of these moms blog on motherhood, but many use their blogs as a sounding board on life. Whether it's not being satisfied in the bedroom or resenting housekeeping, moms have an outlet for their frustrations to be heard.

While the popularity of blogging evolved from the comfort and anonymity users enjoy on the Internet, the motivation has been different for mothers. Blogging was a natural move from populated message boards and chat rooms of the late 90's where moms traded experiences and validated their parenting skills with other moms. Blogging provides moms the ability to go beyond reacting to subject matters, as is the case with message boards, to produce their own thought provoking discussions. It also fills an emotional element for mothers by eliminating the feeling of isolation that can occur when moms are engrossed in their parenting responsibilities. Moms repeatedly tell me that blogging makes them feel like they are not alone in the challenges and experiences they face each day as a mother.

Of all the new media out there today, blogging is the best aligned with the emotional and physical features of a mom. It's a communication vehicle that allows her to share ideas or give her opinion in a format that is easy to use and fits into her busy schedule. It doesn't have strict formats and requires little to no technical knowledge. In fact, regardless of her generation label, she probably already has been trained to blog. What little girl didn't keep a diary or journal at one point in her childhood? It's no wonder that many mom blogs read almost like an open diary of her life as a mom. Best of all, unlike video casts, she doesn't have to worry whether or not she needs more lipstick to engage in blogging. It can be done in pajamas at 2:00 a.m. with a Cosmopolitan sitting next to the computer, or from her mobile phone while sitting at a practice field.

How Marketers Can Use Blogs

For marketers in the Mom Market, blogs can be used in three significant ways. The most obvious way is to share your product and services with mommy bloggers who then spread the word to their audiences as a trusted third party. We will spend most of this chapter discussing the strategy and steps to reaching this goal. Secondly, companies can start their own blog. This allows you to become part of the blogging community and engage audiences in firsthand information about your company. Much has been written on the best practices and case studies of corporate blogs so I'm going to leave the technical instruction to others.

I do have a few suggestions for setting up a blog focused on attracting moms. Consider who within your company will carry your voice in the blogosphere. The most likely candidate should be a female employee with children. She doesn't have to be senior management to be effective. Moms like to follow women with similar lifestyles. If you are a youth shoe company and one of your product designers happens to be a mother of a two-year-old, she may be the better candidate than the male Vice President of Marketing with two teens. My second suggestion is to broaden your topic. In the case of the shoe company, the inclination might be focus the content of the blog on future designs and new product—but blogging is about transparency and reality. It would be far more interesting to mothers to follow your young shoe designer as she balances work and family while chronicling the hundreds of failed designs that got her to this year's new product launch.

It is important to mention the second, and often overlooked, functionality of blogs for marketers—research. It's called "blog mining" and it can provide a rich gathering of consumer insights. Amongst the sharing that is extended in mom blog content is a great deal of personal opinions and reactions by her audience. Imagine how valuable it would be to discover that moms are recommending your product based on a new use or functionality? It could open up new channels of distribution or maximize the effectiveness of your advertising messages. The Blackberry is a good example of this. The device was originally marketed to business professional, specifically managers who wanted to stay in touch with employees. However, it quickly found a presence on the sidelines of soccer fields where moms were syncing up their schedules on their Blackberries.

The use of mom blogs allows marketers to maximize the influence of Mom Mavens. This group, as we know, has long shared information about products, retailers and services with other mothers. Today with the help of the Internet, they can tell thousands, and in some cases millions, of other mothers. I hear more and more from moms that they have added mom blogs to their list of resources when researching a new product, travel destination or retailer. One mother recently described her search for sneakers for her child to me on Mom Talk Radio. "I was unsure about spending $100 on a pair of shoes for my child so I left the store and went home and searched mom blogs to see what other moms thought about the shoes. In the past, I would have gone to Consumer Reports first," she explained. Over the last six months I have heard this scenario repeated by literally thousands of mothers.

Mom blogs have made it easy for the time-starved mom to get the opinions of other mothers—without all the lengthy phone calls or playground conversations. She gets the same information she used to get by jumping out of her car at pick-up time and chatting with other mothers in seconds, and in the convenience of her living room.

The timeless availability of the information is also a huge factor in the popularity of mom blogs. Our shoe-seeking mom might not have the time to ask other moms before 10:00 p.m. and while this may be too late to call a fellow mom on the phone, it's not too late to learn and interact from other mothers online. Perhaps this is why my research shows that over 60% of moms are online from 8:00-10:00 p.m.

Rules of Engagement

Marketers have quickly come to realize the power of mommy bloggers. Two great marketing initiatives rolled into one—viral online marketing and word of mouth influencers. However, it's the "how to use mom bloggers" question that has created so much debate in the blogosphere and even more discussion in corporate America. The undefined rules of engagement with bloggers combined with the tidal wave of influence—both positive and negative—that can be generated by bloggers has forced many companies to make mistakes along the way.

Take the example of an egg company who sent out invitations to mom bloggers to distribute their Easter ideas to other moms. A great idea that, you'll see, was poisoned by poor execution. The mistake that these eager marketers made was to send the Easter pitch to Jewish mom bloggers who were insulted that the marketer didn't take the time to get to know them. The egg company received a great deal of negative reaction throughout the blogosphere. In traditional outreach, I'm sure that the egg company would not send the Easter ideas to the sports editor at a newspaper, but to the appropriate features writer. The same rule applies with mommy bloggers.

Marketers and public relations professionals must take the time to get to know their target audience. It's the first rule of public relations and the first rule of mommy bloggers. Take the time to read several posts on a blog and assure yourself that your pitch fits the style of the blog. Conversely, you also want to make sure that you want your brand represented by the mom blogger you are contacting. Not all mom bloggers are created equal and if you have a problem with the "F" word being used in the same paragraph as your product's name, it is in your best interest to identify the style of the mommy blogger. Even bloggers themselves identify with several distinct sub-segments of mommy bloggers and as a marketer you must consider your choice of partners just as you would in the offline world.

Types of Bloggers

Before we move into our discussion on bloggers, let's briefly go over the types of mom bloggers in today's blogosphere. The largest groups of mommy bloggers are technologically savvy young women who enjoy the experience of sharing information with other moms, chronicling their journeys through motherhood and meeting like-minded women. They blog with pleasant and positive styles. Many have accidentally fallen into a business opportunity that they are enthusiastically attempting to manage.

On the other side of the blogosphere is a group informally called "The Edgy Mom Bloggers." It's not uncommon to read profanity in their blogs. They use the shock factor to drive discussions on their blogs and rise in Technorati ratings. These moms are not afraid to express their opinions or tackle controversial topics such as sex, politics and religion. They are a sub-culture of mommy bloggers that will call out companies for making mistakes in marketing or appearing to be insincere in their pitches. The greatest mistake a marketer can make with these moms is to approach them without taking the time to read their posts. Be warned!

Marketing with mom bloggers is as much on the minds of bloggers as it is with marketers. There have been many discussions recently in the blogosphere among moms as to how to manage the sudden attention they are receiving from companies. I thought it would be beneficial to give you a glimpse into these online conversations.

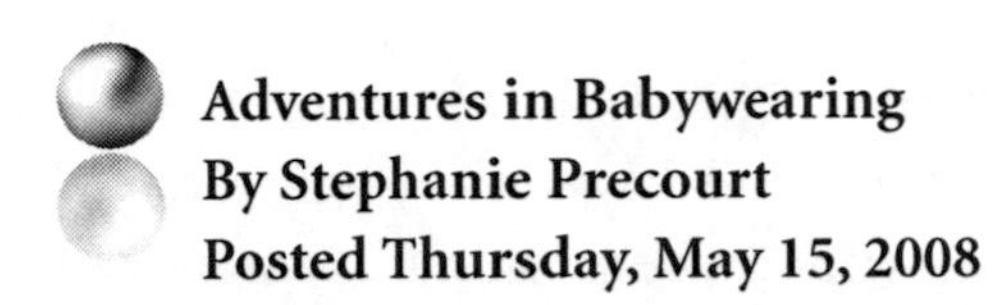

Adventures in Babywearing
By Stephanie Precourt
Posted Thursday, May 15, 2008

This Is Not My Soliloquy

When I first started blogging, I didn't realize that there was really anyone else out there like me, and I never would have thought they'd read what I had to say. I was talking to myself and that was quite alright. Suddenly what I was doing had a title—Mommy Blogger—and I found out my little corner of the Internet was actually in a really, really big neighborhood. A community actually. One that continues to grow and water itself and grow some more. And I love being a part of it. I love that I'm not talking to myself anymore (at least online...).

So I'm a Mommy Blogger and I write about Mommy stuff and life stuff and how much I love my kids and never want them to leave my side and then the next day how I can't wait to get away from them. All the blogs I read, the people that leave comments and my readers—you are all my entourage. I take you with me wherever I go. What a resource to have in any time of need. I can reach out and have an answer in an instant—right at my finger tips. Who needs Google anymore when you've got a Mommy Blogger on your side? We know people. We know lots of people. We're a strong army of knowledge and wit and emotion and power.

We don't have to hang out in real life to care about each other, cry on virtual shoulders, and lift each other up or make each other laugh. From the truly heartfelt and thoughtful gifts like dresses to the congratulatory flowers delivered to my doorstep yesterday from my blogging friend MJ. I feel the love. And it rocks. You rock.

So next week I get to have another little get away. A night out in the big windy city and an opportunity to share my voice as a Mommy

Blogger. Maria Bailey from BSM Media, one of my favorite people, is bringing some companies and bloggers together to help set the record straight at something she's calling Mommy Monologues. How do we want our voice to be heard? How do we want to be approached about product reviews and as a *brand*?

And *do* you want to be approached at all? We're a target market, maybe one of the biggest. Because when someone in my online entourage says something, I listen. And that has more power than any magazine or commercial you're going to fast forward during *The Office* tonight.

But what we have to say is worth something. To each other, we share freely. When companies want to cash in on this "something special" that we all have, we have to wise up. Kick off the Crocs and put on our lipstick and power heels and think this through. We remain in charge of where this is all going. Because without our everyday rants and raves and cries for help, without our openness and honesty and reality, there is no voice.

If you've chosen to rebel against Marketing to Moms and ads and the commercialization of it all, stand your ground. I applaud the bloggers who know that route isn't for them and stick with it. But if you're like me and are open to what professional possibilities lay ahead, I invite you to come along for the ride. Maria has formed a new site called Mom Select (www.momselect.com) for the Mom Bloggers that want to learn more about connecting with companies. This is a great place for you if you aren't sure where to start. And ask me anything—I am always glad to help. There isn't anything I won't share with you. I've gotten where I am today because of other bloggers, too. We're all in this together. It's not a competition. It's not a race to who gets the most comments or has the most touching or funniest post of the day. It's just about being you and being real. There is strength, there is purity, and there is *community* in that.

And so, if I'm being honest, when I said yesterday that we don't have cookies for breakfast or even lunch...well...if I said that this morning, I'd be lying.

Written by Stephanie at Adventures In Babywearing - 11:15 AM

Comments Posted on Stephanie's Blog

To Think is to Create said...
Oh yes, we had no idea what we were getting ourselves into, did we? :) No idea it would ever get a name! I think that moms finally have a way to get their own voice out there, and judging by the traffic all our little blogs get, people are listening! I can't wait to see where all this goes some day. So thrilled to be a part of our community and especially part of your audience. Loved this post!
Written - 11:36 AM

The Flip Flop Mamma! said...
When I started blogging I had never even heard of a blog! I thought "cool! I want one of those!" I'm glad to wear the title "Mommy Blogger" now:) and I love the perks that come along with it!
Written - 11:55 AM

pinks & blues girls said...
This is a great post, Steph. I so agree that all mommy bloggers (I'm boldly including myself in this category...!) are not the same. Some are open to pitches; others are decidedly, um, NOT. :) But we are a force that is not to be ignored. So I think it's great to come up with a way that we can agree works for (most) everyone when it comes to participating in this whole new world we've collectively

carved out for ourselves with the help of each other! Excited to hear the results!! Have an amazing time in the Windy City!
Jane
Written - 12:05 PM

Ellinghouse said...
Steph, Great post! Since the beginning of the year somehow (not really sure) I got into some really great product reviews. It's been so fun, and it's true...if one of my bloggy friends post about something I "just gotta have," I at least look into it, if not get it myself. I thank you (and some of your bloggy friends) for really paving the way. Even if I never get to review another product or if my stats dropped to 10 hits a day, I probably would still blog. Because in my mind, this bloggy habit is cheaper than therapy. And that's what I'm doing, living life and venting about the ups and downs. I don't know what I would do without it.

Who would have known that 2 years ago?
Written - 12:12 PM

Classy Mommy said...
Love your post. Love your positive attitude. Clearly, not all Mommy Bloggers are the same and there are many different opinions on the Marketing & PR pitches hitting our audience. People are different and this makes the world go round—however, I find the negativity out there a drain as I'm all about the harmony and the feel good! So keep up the good work and hopefully you'll be getting more invites where you can hideaway the Crocs to slip on your heels and lipstick to impress!
Written - 12:33 PM

Top Mom Bloggers

As you can see, marketers aren't the only people trying to navigate this new and powerful medium. It's my belief that the best insights are garnered by going directly to the people immersed in the target market. For this reason, I thought it would be most useful for you to hear directly from some of the most popular mom bloggers. In an ironic way, you will be engaging with mom bloggers in the same way that moms do with each other every single day by reading their written word. It's with great pleasure that I introduce you to a few of today's top Mom Bloggers.

Amy Mueller of MumsTheWurd. com

Q. What is the name(s) of your blog(s)?
A. MumsTheWurd, LadyBug and her Blogging Mama, ParentSphere

Q. How long have you been blogging?
A. I've been blogging for just about three years now—since July 2004.

Q. What is your estimated audience size?
A. Between all of my sites, approximately 8,000 unique visitors per month.
Q. How do you describe your blog?
A. I am a blogger who writes about unique products that are hard to find on the web, motherhood and current events.

Q. What motivates you to blog?
A. Life! My passion for "being in-the-know" amongst the who's who of the blogging world.

Q. How often do you post to your blog?

A. I blog every day. I think if you're looking to capture an audience and keep loyal readers, you have to keep your blog current. I would say in that case, posting every day is a must to keep the interest of your audience.

Q. Do you read other blogs? If so, which ones?
A. I try! I like to visit my friend's personal blogs and leave comments. I follow Dooce occasionally. I absolutely adore Antique Mommy (who I stalked at BlogHer '07). Mod*Mom, Design Mom (I like modern things), Zooglobble (kid's music) Can You See the Sunset (mp3 blog), i speak film and Photodoto (photography sites), BlackBerry Cool, RIMarkable, Techmamas (I'm a tech geek).

Q. If you could describe one "emotional" feeling you receive from blogging, what would it be?
A. It would be that some day, my daughter (who is 2 1/2 right now) will be able to look back through my blog(s) and read about everything I did while she was growing up. I want her to be proud of me and really know me. Hopefully it won't come back and bite me.

Q. Marketers are now looking for ways to get into the conversation of mom bloggers and their audiences, how do you feel about that?
A. I think it's wonderful. Marketers can gather so much valuable information when working directly with moms who will test and provide feedback on their items and hopefully the end result will be better, more efficient, more usable products when the finished product hits the shelves.

Q. Do you take paid advertising on your blog?
A. Yes.

Q. What is in your opinion, the best way for a marketer to work with you or other mom bloggers?

A. What has worked for me is direct contact or a personalized email and building a relationship. I don't mind either form of contact. In fact, I have my number listed on my blog. I find that touching base with the "mom blogger" by phone is a very good way to stick out in their mind. Find out what really makes that specific blogger tick. You will run into mom bloggers who are happy to help promote yours or your client's products and those who will bark at you for not taking the time to read their blog and understand the content they provide to their readers.

Q. Do you think you are jeopardizing anything by working with marketers to present new products, services or samples to your audience?

A. No, but I do believe that if the blogger is going to promote products (new or old), they should really do their research. They can't claim to be 100% "green" if they are not recycling, for instance. Their readers will come back and accuse them of being hypocritical. It's important to really know the product (and company) that you're promoting, inside and out. It all really depends on what kind of blog and audience the blogger has.

Q. Where do you think blogging will be 3 years from now?

A. A common household name. Every professional, SAHM (stay-at-home mom) and homeless person will have a blog. Corporations will have blogs, stores will have blogs. Hopefully the President will keep a personal, public blog. It's the new advertising outlet and the newest form of communicating between the people of the world. We're one step closer to telepathic communication.

Q. Were you the kind of girl who kept a diary?

A. You betcha.

Q. Why does blogging appeal to you more than Vlogging, Podcasting, Message Boards or other social media on the Internet?

A. Blogging is a bit more personal and you can control your audience to a certain extent (i.e. comments). You can choose to keep your blog private or make it public. You can "hide." Message Boards are very public. By leaving a comment on a Message Board, you risk dealing with criticism and heated emotions. I think Podcasting makes you seem more vulnerable. People can really "hear" you in every sense of the word when you let your voice and emotions be heard over the air waves. Same goes for Vlogging.

Q. How likely are you to be blogging in 2 years?
A. No question, I'll be all over it.

Jo-Lynne Shane of Musings of a Housewife

Q. What is the name of your blog(s)?
A. Musings of A Housewife, Chic Critique

Q. How long have you been blogging?
A. I have been blogging for two years; I began my blog in March of 2006.

Q. What is your estimated audience size?
A. According to Site Meter, Musings of A Housewife has around 500 visitors per day, which never ceases to amaze and amuse me. Why people want to read about my daily life is beyond me! Chic Critique is my newer venture, and is averaging around 100 visitors a day, I believe.

Q. What motivates you to blog?
A. Well, my teachers always did tell me I talked too much. I guess I've found a way to talk all I want. Whenever anyone gets sick of it, they can just turn me off.

Q. How often would you say you blog?
A. I post on average two times a day.

Q. Do you read other blogs?
A. Yes, I have about 75 blogs I follow on my feed reader. I read a variety of blogs—parenting blogs, fashion blogs, beauty blogs, blogging blogs. I used to try to read the blogs of all those who left regular comments on mine, but that eventually got out of hand.

Q. If you could describe one "emotional" feeling you receive from blogging, what would it be?
A. Satisfaction. Unlike motherhood and housekeeping, there is no one to come along behind me and mess up what I've created. There is something very satisfying about working hard on something and then being able to sit back and enjoy it for more than five minutes (which is about all the time it takes for the house to get messed up after I clean it.)

Q. Marketers are now looking for ways to get into the conversation of mom bloggers and their audiences, how do you feel about that?
A. I love it! I liken mommy bloggers to the talk shows of the 80's and 90's and the radio shows of the 40's and 50's—women sharing tips on running the household, fashion, raising kids, products, etc. It's just another media outlet. Women used to sit down with their morning coffee and watch *The Today Show*. Now they sit down with their morning coffee and read blogs. Everyone has a DVR and can fast-forward through commercials. Advertising on TV isn't as effective as it used to be. So marketers are targeting popular blogs for their advertising. It's the next trend in media outlets.

Q. Do you take paid advertising on your blog?
A. Yes, I do. It's nice to cover the costs of blogging and to earn a few extra pennies for doing something I'd be doing anyway.

Q. What is in your opinion, the best way for a marketer to work with you or other mom bloggers?
A. Blog readers love giveaways. I mean, hello, who doesn't? You can't beat FREE, you know? When marketers ask a blogger to review a product, their publicity multiplies exponentially when they also offer to giveaway a second product to a commenter. Note to marketers: when offering to do a giveaway through a blogger, please offer to ship the product directly to the winners. It is a hardship on the blogger to pay for shipping, particularly if the object is large or heavy.

Q. Do you think you are jeopardizing anything by working with marketers to present new products, services or samples to your audience?
A. Not at all. I don't know what that would jeopardize. I suppose an over abundance of such posts might turn some readers away, but what the readers gain, in my experience, makes up for any loss.

Q. Where do you think blogging will be 3 years from now?
A. I have no clue. The world is changing so fast, and there are hundreds of new bloggers every day. It will be interesting to see if it a passing phase. The social networking sites are the newest buzz around cyberspace, but I don't see them replacing blogs. They are much harder to navigate for information, in my opinion. They are more of a social outlet.

Q. Were you the kind of girl who kept a diary?
A. No! That is the funny thing. I could NEVER keep up with a diary, and when I started my blog, I never imagined I would be this prolific.

Q. Why does blogging appeal to you more than Vlogging, Podcasting, Message Boards or other social media on the Internet?

A. I am not a huge fan of the podcast. I don't like YouTube for the same reason. I am too ~~busy~~ impatient to sit and listen to someone yammer until they spit out what I could read in half the time. Likewise for message boards, it is too time consuming to sift through all the junk to find something I might want to read. My affair with the Internet began with a message board, Babycenter.com, to be exact. I still communicate with some of the moms I "met" on one of their birth boards in 1999! But I've moved on to blogs and social forums.

Q. How likely are you to be blogging in 2 years?
A. I'd say very likely. I hope to grow my audience as well as my sphere of audiences. I am now contributing on two blogs that I do not own, as well as on the three that I do own. One that I own is a multi-author blog, and I hope to add more contributors as it grows. I would also like to expand into other writing markets. I have submitted several magazine articles. We'll see where it goes!

Q. Is there anything else you'd like to say regarding new ways for marketers to reach moms? Do you think they are doing a good job? Do you feel they are insincere in their relationships with moms? What do you want most for marketers to know?
A. I have been pleased with most of the marketers I have worked with. They are generally very personable and responsive. Particularly Natalie of BSM Media!

Christine Koch of Boston Mamas

Q. What is the name of your blog(s)?
A. My primary blog is Boston Mamas, www.bostonmamas.com.
I also blog through my design business Posh Peacock,
http://blog.poshpeacock.com.

Q. How long have you been blogging?

A. I founded Boston Mamas in July 2006 and started the Posh Peacock blog in October 2007.

Q. What is your estimated audience size?

A. For Boston Mamas, around 16,000 unique visits per month and growing, across North America with some international too.

Q. What motivates you to blog?

A. I find it so ironic that before I started blogging, I never read a blog regularly and barely knew what they were about. And then when I first chatted about my idea for Boston Mamas with a tech friend, he immediately suggested blogging as the platform…the rest is (short) history.

On a basic level, part of the motivation for me is that I find the dynamic nature of blogs so captivating. I spent 10 years in academia (I have a Ph.D. in music and brain science), where gratification typically is long-term (and often painful and dragged out); blogging is the complete opposite of that. I have so much content to share through the site and the novelty of the site looking different every day by virtue of new posts still has not worn off for me. And of course, I'm also motivated by my readers—I'm grateful for the kind feedback I have received from readers, colleagues, and marketing/media professionals; it inspires me to continue providing high quality, thoroughly researched content generally (for the non-Boston contingent), as well as a cool niche for moms in the Boston area. The Posh Peacock blog was inevitable given that the business was new and I wanted to be able to remember, years from now, how and when all the different changes (e.g., new designs, retailers, press, shows) took place.

Q. How often would you say you blog? What is the ideal frequency for new postings?

A. At Boston Mamas I typically post twice a day during the weekdays. I used to post on the weekends too, but I only do that now if there's something very pressing to report. I have found that taking the weekend off rejuvenates my writing spirit for the following week. At Posh Peacock I blog as news arises.

Q. Do you read other blogs?
A. I'm slightly embarrassed to say that between my family, Boston Mamas, Posh Peacock, and my other freelance writing and editing contracts I don't have a lot of time to surf. In that limited window, though, I'll admit that I check out People's daily celebrity photos (pure candy), and otherwise a couple other favorites include Universal Hub (for the informative and absurd in Boston) and—due to my obsession with textiles—Sew, Mama, Sew! And I have a cross posting consortium of stylish mommy bloggers that I check in on Friday mornings. If you'd like a closet surfing secret, I'll also confess that back when Michelle Kwan was still skating competitively I used to regularly check in on a fan site of hers; it was before blogs came to fruition but the owner—a veritable content machine!—dated each update in static format.

Q. If you could describe one "emotional" feeling you receive from blogging, what would it be?
A. Gratitude. I'm grateful to have this passionate project after spending a lot of passionless years in science. I'm grateful that people care about what I have to say and have offered so much great feedback about the site.

Q. Marketers are now looking for ways to get into the conversation of mom bloggers and their audiences, how do you feel about that?
A. I think this is a natural trajectory as far as marketers serving the purpose of trying to get products out there, and moms being the ones who control a lot of household spending. But for Boston Mamas, I have firm standards in dealing with *anyone* who wants to get product in front of Boston Mamas' readers. And I think

limiting product review space (since I cover a lot of other topics) and not running advertorials means that the value of a product featured on the site is greater.

Q. Do you take paid advertising on your blog?

A. Yes, via Blogads. And it is very clear on the site where the sponsored ads are.

Q. What is in your opinion, the best way for a marketer to work with you or other mom Bloggers?

A. I think the key is that marketers need to respect the standards that a particular blogger has set for their site. For example, if a marketer is only willing to send a press release for a product, that's fine, but it means that the product will definitely not get air time on Boston Mamas (whereas it might on other sites). We want to actually see, feel, and test drive items, and the requirements stand the same for any retailers, vendors, and marketers. What I don't appreciate is when I take the time to respond to a PR request and very clearly point to our review guidelines and then receive no response but keep getting deluged by repeated requests to feature press releases.

Q. Do you think you are jeopardizing anything by working with marketers to present new products, services or samples to your audience?

A. Not if you believe in the product. But if a blogger is presenting the product in the form of an advertorial (or something similarly paid) then that fact must be absolutely clear. When I first learned about advertorials I was *so* depressed. I wondered how much I had purchased in the past based on what I thought were real features or reviews. And now I look at magazine or online product features with a much more careful eye.

Q. Where do you think blogging will be 2 years from now?

A. Still running strong!

Q. Were you the kind of girl who kept a diary?

A. Yes, for a while. But then when I was in college I read back on one of my old journals from middle school and was appalled by how many brain cells I was burning moping about this or that guy. It was the sort of thing I definitely didn't want anyone to find and publish as my memoirs! And yes, I destroyed the documentation.

Q. Why does blogging appeal to you more than Vlogging, Podcasting, Message Boards or other social media on the Internet?

A. As I said, I was basically new to blogging when I started and it fulfills my media needs currently. Some people have asked me about the possibility of starting message boards on Boston Mamas, which seems like it could be a good thing, but then I have visions of my life revolving around moderating comments and/or—as I'm afraid I have heard about— posting faux messages to make it look like the boards are active. Dreadful.

Q. How likely are you to be blogging in 2 years?

A. I'll be there unless I'm trapped under something really heavy!

Q. Is there anything else you'd like to say regarding new ways for marketers to reach moms? Do you think they are doing a good job? Do you feel they are insincere in their relationships with moms? What do you want most for marketers to know?

A. As I said, respect the standards set by the site. And one piece of advice is to actually spend the extra few minutes on the site to see who is behind it (the About page is usually easy to find). I'm actually very good about trying to respond to all inquiries, but in a flurry of clean up, I'm much more likely to discard form letter emails addressed to "Hello" or "Hello important mom blogger" than "Hi Christine." I know marketers are canvassing a lot of people, but if they want to make a connection with the mom blogger(s) behind a site, they need to start off by acknowledging that a real human lives behind the keyboard.

Janice Croze and Susan Carraretto of 5 Minutes for Mom

Q. How long have you been blogging?
A. Two years. Launched our blog in March 2006

Q. What is your estimated audience size?
A. 60,000 monthly visitors

Q. What motivates you to blog?
A. As work at home moms who own two online stores, we know how hard it is to start and run online businesses AND to be stay at home moms. We want to provide a fun, relevant blog that helps moms support and find one another.

Q. How often would you say you blog? What is the ideal frequency for new postings?
A. We post approximately twice a day—sometimes more. Ideally a blog should be updated at least daily.

Q. Do you read other blogs? If so, which ones.
A. Yes—but that is a hard question to answer! There are tons that I would recommend. For this reason, our blog includes detailed blog directories to help moms find other blogs that interest them. One of my favorite blogs to read is http://boomama.net/.

Q. If you could describe one "emotional" feeling you receive from blogging, what would it be?
A. The best feeling I receive from blogging is feeling connected with other moms. It is wonderful to be part of this community of moms supporting each other and sharing the fun of life through our blogs and our businesses.

Q. Marketers are now looking for ways to get into the conversation of mom bloggers and their audiences, how do you feel about that?

A. I am all for it. Marketing makes the world go round! We need marketers and they need us. Moms do want to know about great products—they just want to be informed in a respectful and fun way.

Q. Do you take paid advertising on your blog?
A. Yes.

Q. What is in your opinion, the best way for a marketer to work with you or other mom bloggers?
A. We have fabulous relationships with the companies and PR companies that we work with. They are respectful and accommodating and work hard to ensure that our posts are meaningful and fun for our readers. As I said, I don't think readers mind advertising, but they want to be informed in a respectful and fun way. For example, giveaways are a great way to help make advertising acceptable on our blogs. Readers will read a post that is advertising to them if it is written in a fun and unique style and if they can get something out of it—like the chance to win something. Everyone loves free stuff. It is like we are "paying" our readers for their time—and they appreciate that gesture.

Q. Do you think you are jeopardizing anything by working with marketers to present new products, services or samples to your audience?
A. It is tricky territory. Bloggers need to be careful to respect their readers' time and not just advertise to them. Readers are at our sites to be informed, entertained and in our community. We can't publish too many "ad" posts on our blogs. As well, those marketing posts that we do publish must be exciting, interesting and draw in readers. We can't allow "ad" posts to hurt our blogs. We need to keep in mind that marketing must always be full of benefits for our readers or else we all— marketers and bloggers— risk alienating our readers. Having said that, for a blog like ours

who's fundamental purpose is to provide resources for readers and links to great blogs, stores, products etc., we are not at the same risk as regular mom bloggers. Readers expect to find information about products at our site. They come to us expecting contests and reviews. They want them. So we can publish more of these posts than a regular parenting blog can. But our readers do want exciting, fun posts where they have the chance to win something great or get something for their time. We have to respect our readers and make it worth their while to read our blog.

Q. Where do you think blogging will be 2 years from now?

A. I think it will just keep growing. Blogging is interactive and people want to be involved in the conversation. It is an incredible avenue for community. I also think advertising will be an integral part of it. Bloggers who are spending a significant quantity of time on their blogs need to be financially compensated. Hopefully readers will understand and respect that. When we read magazines or watch TV, we understand that someone has to foot the bill. Bloggers also need to cover their costs. When I read blogs, it doesn't bother me at all to see advertisements. I want that blogger to be compensated.

Q. Were you the kind of girl who kept a diary?

A. Yes—I imagine most of us bloggers were.

Q. Why does blogging appeal to you more than Vlogging, Podcasting, Message Boards or other social media on the Internet?

A. Blogging is a form of creative writing. Even if it is a product review, each one of my posts is a piece that fulfills my need to create. I prefer to communicate in the written word that is crafted and edited, unlike the faster "chatting" methods of message boards and the verbal forms of podcasting and vlogging.

Q. How likely are you to be blogging in 2 years?
A. I will still be blogging in 2 years. Unless, of course, I am dead. Here is hoping I am still blogging.

Q. Is there anything else you'd like to say regarding new ways for marketers to reach moms? Do you think they are doing a good job? Do you feel they are insincere in their relationships with moms? What do you want most for marketers to know?
A. The PR companies we are working with are doing a great job and have treated us, and our readers, well. I want marketers to know that blogging is an effective form of marketing. The conversations we are having online directly affect the way we shop. Therefore, I hope in the future companies who are planning marketing campaigns allocate a more significant portion of resources for the PR companies to work with this new media. It is a cost effective form of marketing and shouldn't be overlooked.

Stephanie Precourt of Adventures in Babywearing

Q. Tell us about your blog(s)?
A. Adventures In Babywearing. I also blog for NWIparent magazine *Close To Home* and am a contributor at Mama Speaks and the Chicago Moms Blog.

Q. How long have you been blogging? What was the date you began your blog?
A. I have been blogging a little over 2 years now. I started with my Adventures In Babywearing blog on December 30, 2005.

Q. What is your estimated audience size?
A. Currently it's about an average of 450 visits per day. There have been times it's much more or less.

Q. What motivates you to blog?

A. If something is bothering me or needs awareness, I am excited to blog about it. I like to get information to people and also stir things up without too much controversy. Having so many readers motivates me to keep things fresh daily so that they always have something new to read. Also having a specific journey to write about (like my new pregnancy) provides me with many new things to blog about and for others to follow along with!

Q. How often would you say you blog? What is the ideal frequency for new postings?
A. I usually blog at least once a day, sometimes more.

Q. Do you read other blogs? If so, which ones?
A. I do read lots of blogs—mainly from other bloggers that I've become friends with. My favorite blog is Lifenut and I actually got to meet with Gretchen in person last summer while I was on vacation.

Q. If you could describe one "emotional" feeling you receive from blogging, what would it be?
A. Release.

Q. Marketers are now looking for ways to get into the conversation of mom bloggers and their audiences, how do you feel about that?
A. I really like knowing that what we say matters and that our voice is being heard. Too often we see commercials and don't know for sure how the product truthfully is in real life, so we all like to tell each other what we really think! Or we have great ideas of our own. Hearing that a product is great from another friend will make me more likely to buy it than if I just saw an ad for it.

Q. Do you take paid advertising on your blog?
A. Yes, but only if it is something that I feel represents me.

Q. What is in your opinion, the best way for a marketer to work with you or other mom bloggers?

A. We love doing giveaways and product reviews that aren't too complicated or require too many hoops to jump through. We love free products and being able to post openly about them. Give us proper compensation—it takes time to learn about the product, find the time to blog, reach our readers, and post about the product.

Q. Do you think you are jeopardizing anything by working with marketers to present new products, services or samples to your audience?

A. Sometimes I've felt that too many reviews can distract from sincere blogging and writing, so I try to keep my reviews and product mentions at my Close To Home blog or at Mama Speaks. A lot of bloggers are keeping review blogs and personal blogs separate now, but still linking to them so that their audience can benefit from great product reviews and giveaways.

Q. Where do you think blogging will be 3 years from now?

A. I think in the past year there has been an explosion in blog land—and in three years for certain everyone will have a blog. Heck—even my own mom has a blog now! Blogging will be where people will turn to for information and daily reads rather than fiction books, email, and newspapers first, I predict.

Q. Were you the kind of girl who kept a diary?

A. I did—I was more of a journal keeper rather than telling secrets to a diary.

Q. Why does blogging appeal to you more than Vlogging, Podcasting, Message Boards or other social media on the Internet?

A. I like blogging because it can be one-sided. I can post and then don't have to necessarily respond back in any way—and the words are mine and not written to any one specific person. I feel that blogging is more personal and controlled. I can make it whatever I want.

Q. How likely are you to be blogging in 2 years?
A. Very likely.

Q. Is there anything else you'd like to say regarding new ways for marketers to reach moms? Do you think they are doing a good job? Do you feel they are insincere in their relationships with moms? What do you want most for marketers to know?
A. I think so far all has gone well with marketers reaching moms. They do need to be careful to have knowledge of the blogger they are contacting and try not to blanket too many blogs at one time with the same promotion or product review. After a while you just turn it off if it doesn't seem unique. Get their name & blog right! Many times I'm contacted via form letters where the wrong blog is entered, etc. Treat us professionally just like we are in regular business—no matter if we're just "Moms!" We have some of the most powerful buying voices out there!

Collaborating With Bloggers

Now that you've heard the voices of real mommy bloggers, how can you go about working with them? There are many ways for marketers to work with mom bloggers. The team at BSM Media has given some of the most commonly used initiatives marketing terms such as Mom Blogger Mixers℠ and Online Media Tours℠. Assuming that you have taken the time to get to know the right mom bloggers to include in your efforts, these two initiatives are quite effective.

Mommy Blogger Mixers℠

A Mommy Blogger Mixer is a social event that gathers numerous bloggers together. The event gives you the opportunity to immerse multiple mom bloggers in your brand or product experience while providing attendees a platform to socialize with other mom bloggers. We have held numerous Mommy Blogger Mixers throughout the United States. Each is themed with the product or brand in mind. There are certain markets with high concentrations of mom bloggers which lend themselves best for these events. New York, Chicago, Washington DC, San Diego, LA, Boston and Atlanta are all prime markets for Mommy Blogger Mixers. Working with groups such as Chicago Mom Bloggers or Parents Blog Network can help in ensuring attendance. Don't forget that these women are moms so events that include their families are popular in addition to mom-only activities.

There are many variations of Mommy Blogger Mixers sponsored by companies. Lands' End sponsored a VIP Blogger preview for its Fall clothing line, HP partnered with DreamWorks to create Kung Fu Panda Parties and Macys hosted in-store fashion shows for mom bloggers. The goal of marketers hosting these events is to ultimately have moms posting on their blogs about the products or the brand they experience.

Although many companies use their public relations groups or agencies to execute mixers, I always think it's a good idea for company representatives to attend and get to know attendees on a one-to-one basis. Remember blogging is all about relationships and two-way dialogue. There is no better way to develop a relationship with a mom than to share a glass of wine with her or interact with her around a play mat. I would encourage you to invite several moms from your organization. This demonstrates a human side to your brand and company and will go a long way in the moms feeling your transparency.

Like all event planning, the devil is in the details. Ensure good attendance by scheduling the event during a time that most moms can attend. Conducting an event for moms during naptime or a moms-only event on busy Saturday afternoons is not recommended. Put yourself into the life of a mom to determine the best time for your event. As always, also make certain your event does not fall on religious holiday or family honored holidays like Father's Day. I had a company who called me excited about a great concept for Mother's Day. They would invite mom bloggers to a Mom's only spa retreat over Mother's Day weekend. Unfortunately, I had to douse their enthusiasm and inform them that moms like to be with their families on Mother's Day. It's never a good idea to schedule events for moms on Mother's Day; companies just can't compete with good old-fashioned breakfast in bed and family time. Finally, as we have mentioned many times throughout this chapter, make certain your invite list includes the RIGHT mom bloggers. Get to know them and ensure that the event theme, brand experience and corporate culture are in sync.

Online Media Tours℠

Online Media Tours are another way to approach mom bloggers. Think traditional public relations meets mom bloggers without the journalistic standards and a greater willingness on the side of the pitchee to get to know you. This is a good strategy for marketing professionals who are looking for multiple posts and mentions by mom bloggers about their product. I caution you, however, that the product review category among bloggers is getting crowded and may be eroding the truly valuable aspects of word of mom. As more and more companies are shipping product samples to mommy bloggers and more bloggers are doing reviews, readers are becoming selective about where and from whom they take product recommendations. Many mom bloggers have established review sites and sweepstake sites in order to work with companies and maintain the integrity of their personal blog. I do not discourage brands from conducting Online Media Tours, I merely suggest getting creative in the approach your company uses to engage moms. For instance, many bloggers are seeking traffic to their site so consider programs that will help the mom reach her goal.

Recently, a company whose goal was to develop a deeper relationship with mom bloggers decided to award scholarships to the BlogHer conference. Instead of just giving each mom one scholarship, they gave each two scholarships. One for them to keep and one for them to use in any promotional manner they elected to reach the goals of their site. Many ran contests to drive traffic to their blogs. The best part of this strategy was that, in normal instances, the company may have received a post or two thanking them for the scholarship, but by allowing the mom blogger to be creative they received recognition throughout the life of the blogger's sweepstake and promotions. Engaging mom bloggers in new and original ways will go a long way to raising your brand above all the others competing for their attention.

Engaging mom bloggers is an effective strategy to create a buzz among moms and, as other technologies emerge, I believe it will continue to grow. Blogging leverages a natural behavior of moms that can predict the longevity of a marketing strategy. Journaling is an activity that teen girls enjoy while growing up and when they become moms blogging allows them to chronicle their parenting journey. It doesn't require as much time investment as scrapbooking and gives them opportunity to socialize with other like-minded mothers. Even as moms move to video, the fulfillment women receive from writing will always be present. While some will gravitate to the next hot technology and others will express their thoughts via multiple mediums, many moms will remain comfortable expressing themselves behind the comfort of the written word. As one mom recently said online in a post, "I don't have to put on make-up to blog."

Well into the future, I see moms continuing to use blogs as a way to connect with their family, friends and children. Marketers can share in the connections mothers are making among their peers by developing creative ways to enhance their relationships with each other as well as with brands, products and services. Whether you elect to conduct sweepstakes, mixers or product reviews with mom bloggers, it is most important to select your partners much as you would select your friends in the physical world. Get to know their styles, values and goals and work together to develop a win-win relationship for both of you. Once you have opened a dialogue with mom bloggers that is based on transparency and trust, they will reward your company with exposure to millions of moms online. Ultimately you will be marketing your product with the help of a trusted third-party influencer—the mom blogger.

Companies who want to keep up with the speed of mom will want to consider podcasting and radio as means to tagging along with her.

Broadcasting Your Message Via Podcast & Radio

CHAPTER

FEW MEDIUMS KEEP UP WITH THE PACE OF A BUSY MOM AS EFFECTIVELY as radio and podcasts. This is why it always amazed me that the radio industry has been so slow to adopt mom-oriented programming, or at least women-driven programming. Some may argue that traditional radio doesn't belong in a book focused on new technologies, but I believe that, as old as radio is as a technology, it has not been used to its full potential in recent years. In fact, I would debate that it has not been used to attract women consumers effectively since the days of Proctor and Gamble's sponsorship of radio soaps. My goal in including a discussion of radio in this chapter is to give you tips on maximizing your radio dollars if you are currently spending them on the airwaves and provide you with some ideas on new ways to approach this old medium. Additionally, many of the strategies and tactics of radio and podcasts are similar in nature. In fact, the two forms of communication can be used in tandem to create a very effective means of communication with moms.

In 1992, I learned about the lack of mom programming on traditional radio firsthand when I caught myself singing Barney songs out loud in my minivan when I was driving alone. Not a single child in the car and I'm jamming along to "I love you, you love me." I did what every embarrassed adult would do and quickly popped the CD out of the stereo and turned the radio on. As I flipped from station to station I found only sports talk and politics—all content largely focusing on men. It was at this moment that two thoughts collided in my minivan. There was no mom-directed content on my radio, yet I spent more hours in that car than any other member of my family. This just didn't make sense in 1992 and it makes even less sense in 2008, when companies are even more cautious in the ways they spend their marketing dollars.

Soon after this "ah-ha" moment as Oprah calls them, one of my clients approached me with the challenge of creating a marketing initiative for them that would "raise them above the clutter." Radio was my answer; largely because no one else was doing it—neither in their market nor anywhere else. The second reason for my recommendation was the built-in content I knew they could provide on radio. The client was a Children's Hospital and I knew that, not only would radio provide them access to moms in an "uncluttered" environment, i.e. their car, but it would also provide the hospital a platform to showcase their knowledgeable staff. Hence, the birth of Mom Talk Radio.

The one-hour show began airing in 1999 on a local AM station. The format was designed to provide moms with answers and solutions to everyday questions and showcase their doctors, dieticians, social workers and birthing experts. It was the perfect combination of built-in guests and an audience eager for answers. There was, however, one unexpected event that even I had not foreseen—my client's request for me to host the show. They thought it was important for a non-hospital employee to lead the show to project a sense of objectivity and since I had an existing following of moms from my prior career in their market, I was their choice. Armed with no experience and a script, I entered the radio studio as the host of Mom Talk Radio. I quickly learned that talk radio is best executed as just that, talk radio and not scripted dialogue.

The show's audience rapidly grew and in a market with five other children's hospitals, our client claimed a marketing tool that spoke to moms in a whole new way. Surprisingly, Mom Talk Radio was an effective internal marketing program as well. Doctors and other professionals lined up for the opportunity to publicize their practices to the chief household healthcare officer—moms. Nine years after the first Mom Talk Radio show the program still airs in South Florida, and is now nationally syndicated and heard around the globe as a podcast. Additionally, it has been recognized by *USA Today* and *Parents* magazine as one of the best radio shows for moms. The content of Mom Talk Radio is now also delivered in several different formats, from a full hour terrestrial radio program to Mom Talk Minutes on branded websites. We will discuss the expanded distribution opportunities of radio later in this chapter. As a marketer trying to connect with moms, you are probably wondering why the idea hasn't caught on in more places. The answers rest on the shoulders of radio networks, program directors and you, the advertiser.

In 2000, I had the chance to speak with Gabe Hobbs, who at the time was the head of programming for Clear Channel. It was a month after *Talkers Magazine*, a trade publication, titled him, "The Most Powerful Man in Radio." One could only imagine the excitement I felt to have the opportunity to pitch mom programming to the most influential man in radio. He patiently listened to statistics, strategy, consumer demographics and concepts but when it was all over he simply responded, "Radio for moms just isn't sexy." I sat dumbfounded. Of course it wasn't sexy! It wasn't meant to be. It was meant to be interactive, engaging and informative. It was meant to fill a void in the market and it was meant to give advertisers access to the most powerful consumer group in the US. It wasn't meant to be sexy unless we had Dr. Ruth on as a guest.

He went on to explain that, in the current radio industry, there was nowhere for him to put Mom Talk or any other mom oriented show in his line-up. I recall him using this example. "Where would I put it? Between Howard Stern and Sports Talk? I'd lose my listeners if suddenly the discussion went from politics to potty training." He had a point, but his explanation only validated the current state of affairs that still exists today. Programming directors see their audiences as predominately male, which lead us to you, the advertiser. Radio will remain as it is today until advertisers spending thousands of dollars on drive-time schedules demand more targeted reach. Unless programming directors feel the financial impact of female consumerism in a positive or negative way they will continue in the status quo manner.

Advertisers should no longer settle for "Women 25-45" as an adequate demographic in which to capture moms. It's like throwing a cast net into the river and hoping you pull out one or two edible fish. It's just not the most effective way to spend your limited advertising dollars. However, having described an industry that is less than mom-friendly in content, radio remains one of my favorite means of marketing to moms. Let me explain why.

Why Radio?

The best part about radio is that moms like radio and have many hours of access to it. It is estimated that more than 60% of all radio listeners are female and over 50% of all listening takes place in a car.[1]

---

1 Surface Transportation Policy Project, *High Mileage Moms* (Washington, D.C.: Surface Transportation Policy Project, May 6, 1999)

This is where you get to apply what you know about a mother's behavior to your advertising plan. A large part of a mother's work week is spent shuttling her children to day care, school, and after-school activities or running errands in her car. According to the Surface Transportation Policy Project, a coalition of organizations interested in transportation policy, single mothers spend 75 minutes a day driving, while married women with children drive 66 minutes a day. Eight out of ten women are radio listeners.[2] The best part about the reach of radio is that moms can listen to radio at work, in the car, at home, or at the park. Moms have access to radio virtually anywhere they go during the day.

Radio presents two marketing opportunities. First, radio provides basic spot advertising, which can be 15 to 60 seconds in length. Rates to advertise are based in the time slot and number of minutes you purchase. Every market is different, but radio airtime can run from $25 to $100 a minute in midsize markets. Most marketers include a catchy slogan, mention the company's name at least three times, and offer a bargain to the consumer they just can't wait to get.

Radio allows you to have short lead times in your advertising message, which is perfect for special announcements. By including urgent marketing messages such as "limited time only," "act now," or "today only," you are able to see immediate results. Radio allows you to target groups of mothers with similar musical tastes or interests. When targeting mothers, it is wise to select music genres that appeal to a conservative market, such as easy listening, Top 40, Christian or country music. When buying radio ads, ask about the value-added elements. Often radio stations will kick in tickets to local events, disc jockey appearance at your place of business, or concert tickets to use for consumer promotions.

2 Ibid

Radio Disney

Don't discount purchasing advertising space on Radio Disney just because its programming is focused on children. Remember, in most instances, a mom is driving the car in which Radio Disney is blaring. A few years ago, while doing research on moms and radio, I asked mothers specifically about their thoughts on Radio Disney. Many media buyers would be surprised to learn that moms are listening as well. For a long time Radio Disney was plagued with the inability to offer ratings to weary media buyers who saw it only as an end of the dial station with limited reach. Fortunately, Radio Disney is now rated by Arbitron and can offer skeptical media consultants the numbers they need to sell clients into this space.

Brokered Space

Brokered time is blocks of airtime available for purchase from radio stations, which allow you the ability to air exclusive programming you've designed to reach your market. Brokered time allows marketers to produce their own programming. The growing popularity in talk radio makes specialized programming a new horizon for marketers who dare to be adventurous and creative.

Brokered time is how we originally launched the Mom Talk Radio show. What makes "Mom Talk Radio" successful as a marketing tool is that it positions the hospital as a trusted resource for mothers in the community. Week after week, it is a place for mothers to turn for answers. In addition, the hospital received the benefit of publicity by launching the first radio show for moms in South Florida. Their sponsorship of this show resulted in newspaper articles in *South Florida Parenting* magazine, *Woman's Day*, the *Miami Herald*, and the *Sun-Sentinel.* We also used the show to announce other hospital events, thus saving the money it would cost to purchase radio ads. We have used advertising space on Mom Talk Radio to entice donors to the hospital's foundation. Donors receive 30-second spots in return for their donations. Mom Talk Radio was an out-of-the-box concept that has paid off in many ways for its sponsor.

Brokered time radio programming can help you carve out a niche and position yourself as an expert and resource in the eyes of your target market. There are easily enough niche programming ideas directed to women that you could have an entire network devoted to women's programming. Since mothers assume so many roles, programming directed at their roles will certainly attract attention. With the emergence of podcasting, the use of radio becomes an even more exciting and effective means of marketing to moms.

Podcasting

The birth of podcasting took place in the mid 90's and rapidly became popular with iPod users. Ironically the growth of online radio shows or podcasts was not fueled by Howard Stern wannabes, but by business owners. Entrepreneurs used podcasts to establish a dialogue and relationship with potential customers. Many of these innovative marketers were momprenuers. Technology-savvy Generation Xers who saw podcasting as not only a way to market their business, but to socialize with other moms. Soon shows such as Mommycast, WAHM Radio and Manic Mommies sprang up online, giving moms the radio programming they missed on terrestrial radio.

For results-oriented marketers, the evolution of podcasting has been a dream come true. It offers many features that terrestrial radio has not been able to deliver. The first is the ability to track listeners. Software such as Libysyn, www.libysyn.com, tracks the actual number of listeners, downloads and audience size. Secondly, it delivers a marketing message with the shelf life of print. Once a podcast is archived online, a single recorded marketing message can be heard for days, months and even years. Through Libysyn, I see moms download segments of Mom Talk that were taped three or four years ago. Although this is not good for marketers promoting a short term contest, it is a wonderful tool for a company attempting to build their brand.

Next, podcasts allow advertisers to gain broad distribution of their marketing message without the challenge of buying large networks or fragmented markets. As a delivery channel, podcasting is the perfect mechanism. It can travel with Mom as she pushes her stroller via her iPod or it can be played in her car via the same technology. In her home or office, she can listen online and, best of all, she can select the time. Finally, the aspect of podcasting that I like most is that you are delivering your marketing messages to a consumer who has selected to receive it. You aren't casting a net. Instead, the mom is coming to you and making the effort to engage in a dialogue with you. Nothing can be as powerful as a mom saying to a brand, "I'm listening."

After enjoying rapid growth and popularity, podcasting lost its place in the spotlight to satellite radio fairly early in the new millennium. In fact, any time I mention the potential of podcasting to marketers, the topic of satellite radio follows. I believe the perception that satellite radio is more popular than online radio is fueled by the large marketing budget of XM radio. As consumers, we hear more about satellite radio because of the ad dollars behind it and its accessibility in cars, hotels and gyms. However, the number of women listening to satellite radio is less than 20% of its overall audience, according to a former executive I interviewed last year. He explained to me confidentially that signing Oprah and Martha Stewart to the programming was an attempt to grow female subscribers. I've yet to find numbers to validate its success. For this reason and the lack of ability for mid-size, small and start-up companies to tap the potential of satellite radio, I will not cover it in greater detail. Podcasting has remained a constant among changing and emerging technologies. Although I predict that video will trump it in audience size by 2009, I believe it remains a viable and effective way to connect with moms.

One of the earliest adapters of mom podcasters were work-at-home moms. Podcasting allowed them to share their expertise and attract potential clients. By hearing your voice, potentials customers feel like they know you. As with all mediums designed to attract moms, it should not be an entire commercial for your company, but rather content with solutions for their everyday lives. One company that has utilized podcasting to connect with moms is Whirlpool. As I discussed, they broadcast a show called The American Family which focuses on family issues. The show is hosted by an enthusiastic Vice President of Marketing who shares her passion for family with her audience. It is available on their website, but also puts the appliance giant in an unexpected place, in the directory of iTunes, Podcast Directory and Podcast Alley. It takes their brand to moms through a new channel of communication, and delivers their message through content that is timely and relevant.

As with mommy bloggers and moms who run social networking sites, I feel that the best way for my readers to understand mom podcasters is to listen in to some conversations with them. Here are some insights from the top mom podcasters recording today:

Top Mom Podcasters

Marie and Sara of ChitChatMoms

Q. How long have you been podcasting or broadcasting?
A. ChitChatMoms posted our first show on 8/31/2006.

Q. What is your estimated audience size?

A. Based on our reporting software, we have on average 600 downloads per episode. This can vary from month to month, however historically we feel 600 is fairly accurate.

Q. What motivates you to record your show?

A. I think we would still record our show if we had one listener or 1000 listeners. It's a chance for us to get together as friends and have a good "girl-talk" session. It's a chance to vent and release, a chance to brag, a chance to reach out and ask for help or to give advice to those who have asked for help. Through contact with our listeners, we've found that they like our show because we're "friends" and we're "real." Sometimes when you need a friend at 2:00 in the morning, it's nice to be able to hit play on your iPod or mp3 player and have one right in your ear, without the guilt of feeling like you're disturbing someone that you would call on the phone.

Q. How often do you tape/broadcast your show?

A. We try to get together about once every two weeks to do a show. It can vary depending on our daily schedules, but that's the goal.

Q. If you could describe one "emotional" feeling you receive from podcasting/broadcasting, what would it be?

A. Validation would be number one. When you get an e-mail from a listener that says she listened to your show and has felt or done the same thing you have, it makes you feel like a normal mom, no matter how crazy the action or feeling was at the time.

Q. Marketers are now looking for ways to get into the conversation of mom influencers like yourself and into podcasts. How do you feel about that?

A. I'm okay with it to an extent. Some podcasters would really like to promote products that they feel strongly about. As a parent, you have a tendency to want to share the news with the world when you find something that works really well for you and your kids.

Sometimes you can clue someone else in on something great that way. But being paid to just talk about a product that you haven't tried isn't something I really believe in.

Q. Do you take paid advertising on your show? If so, what kind of packages do you offer?

A. We haven't been paid to advertise anything as of yet. We would entertain the idea of a paid sponsorship if it were a product or service that we felt would benefit our listeners in some way.

Q. What is in your opinion, the best way for a marketer to work with you or other mom podcasters?

A. I think the first step would be directly contacting the podcast to see if they would be interested in marketing the product or service. Then the podcasters and the marketer could work out an arrangement that would fit the individual show as well as the product.

Q. Do you think you are jeopardizing anything by working with marketers to present new product, services or samples to your audience?

A. Some people would probably feel that we would be "selling-out." But, if we can continue to produce the same content that engaged their interest in the show, as well as providing them with possible discounts or introducing them to products that may benefit them in some way, I don't see a problem with it.

Q. Where do you think podcasting will be three years from now?

A. I think the media of podcasting is just in its infancy. There is so much potential for podcasters as well as those who listen to podcasts, that it's impossible to predict where it will be in one year, let alone three. Technology moves fast.

Q. Did you have any formal training in broadcasting or radio?

A. None whatsoever. I ask for help from other podcasters and IT friends when I hit a roadblock. Most of the time you can find someone who has hit the same roadblock and they can talk you through it.

Q. How did you learn how to podcast?
A. I started listening to podcasts about six months before actually producing them. I stumbled on a comedy show that a co-worker was producing. I really liked how candid he could be on a podcast, and decided that I wanted to give it a shot.

Q. Why does podcasting appeal to you more than Vlogging, blogging, Message Board posts or other social media on the Internet?
A. I think I like podcasting because it allows people to "hear" you rather than "read" you. We still blog on our website just to provide our listeners with more variety and more daily content. But my favorite way is just to sit and talk like we're all having a drink together. It's more relaxing. Plus, you don't get to backspace when you're podcasting. Once you say it, it's out…unless you want to go back and edit everything. Our show is more raw.

Q. What is the next media you want to venture into?
A. I'm not exactly sure I would want to try doing video podcasting. I'd feel obligated to do my hair and makeup then. The great thing about podcasting is that you don't have to get all gussied up to do it. It's something that you can do when the mood hits that you don't really have to prepare much for.

Erin Kane and Kristin Brandt of Manic Mommies

Q. How long have you been podcasting or broadcasting?
A. Manic Mommies podcast was born on July 10, 2005

Q. What is your estimated audience size?
A. 5,000 downloads per week of most recent episode
35,000 per month (includes back episodes)

Q. What motivates you to record your show?
A. The feedback we get from our listeners. It's tremendous. Many say it's the first show they've listened to that validated their hard work and made them feel it was okay to not be perfect all the time.

Q. How often do you tape/broadcast your show?
A. Every week

Q. If you could describe one "emotional" feeling you receive from podcasting/broadcasting, what would it be?
A. Validation

Q. Marketers are now looking for ways to get into the conversation of mom influencers like you and into podcasts. How do you feel about that?
A. Smart companies recognize the value of honest conversation and can be supportive without being intrusive. Our listeners are super savvy. They recognize a veiled advertisement when they hear one. We will only accept sponsorship from companies we believe in and we will not risk our credibility or relationship with our listeners for corporate money. That has meant some lost opportunities, but we work hard with every potential sponsor to figure out a way to integrate their brand without compromising our content.

Q. Do you take paid advertising on your show? If so, what kinds of packages do you offer?
A. We do accept paid advertisements, on our website and on our podcast. A monthly sponsorship is around $7,500 and includes web, podcast, email newsletter, etc.

Q. What is in your opinion, the best way for a marketer to work with you or other mom podcasters?

A. When we worked with General Motors on our Manic Mommies Escape, they were very careful to not to push their agenda on us. We told them what we needed: courtesy transport from the airport to the hotel and shuttles to the off-site tourist destinations. They provided both and brought in some new model cars for moms to test drive on a pre-determined sight-seeing route. This allowed the moms to see an area of Newport they would not have seen on their own (without a car). They also provided courtesy drivers who did not at any time try to sell the car. The moms were so appreciative of not being the driver for once. It was amazing to come out of the yoga studio and see four SUVs lined up with drivers waiting to take us for coffee and back to the hotel.

Q. Do you think you are jeopardizing anything by working with marketers to present new product, services or samples to your audience?

A. We do a number of giveaways on our show and moms seem to appreciate the opportunity to win stuff. Especially hard-to-get prizes like the Nintendo Wii.

Q. Where do you think podcasting will be three years from now?

A. We hope it will become more mainstream. We still hear from many friends that they can't listen because they don't have an iPod. They don't realize they can listen directly from our website.

Q. Did you have any formal training in broadcasting or radio?

A. Erin worked for several years at WGBH, the public television and radio station in Boston. Kristin is in advertising and has managed the production of commercial spots.

Q. How did you learn how to podcast?

A. We taught ourselves using GarageBand on Mac platform.

Q. Why does podcasting appeal to you more than Vlogging, blogging, Message board posts or other social media on the Internet?

A. We are bloggers, too. But podcasting feels more like a real conversation between friends because you hear our actual voices. Also, the portability of the medium makes it a plus for trying to reach busy moms on the go.

Q. What is the next media you want to venture into?

A. We've recently produced some video spots but in all honesty, we prefer not having to do hair and makeup. We can podcast in our pajamas. And drink wine while we're doing it!

Dave and Heather Delaney from Two Boobs and a Baby

Q. How long have you been podcasting or broadcasting?

A. We began in October 2005.

Q. What is your estimated audience size?

A. 10,000 +

Q. What motivates you to record your show?

A. A chance to sit down together and laugh about our family antics. Plus, listener feedback always gets us pumped to record another episode.

Q. How often do you tape/broadcast your show?

A. We try to record once every two weeks.

Q. If you could describe one "emotional" feeling you receive from podcasting/broadcasting, what would it be?

A. Pride. We're proud to produce a show that people enjoy listening to. We're also proud to be the parents of two incredible kids.

Q. Marketers are now looking for ways to get into the conversation of mom influencers like yourself and into podcasts. How do you feel about that?

A. As a professional New Media Marketer, Dave is thrilled with it. We feel that if we can make enough money from our podcast to work from home, it would enable us to spend more time together as a family. However, money is not why we podcast.

Q. What is in your opinion, the best way for a marketer to work with you or other mom podcasters?

A. The best way would be to determine how our show can help you boost your brand's awareness. We have demographics of our audience available to marketers, so determining if our reach is appropriate would be key.

Q. Do you think you are jeopardizing anything by working with marketers to present new product, services or samples to your audience?

A. No, as long as we both agree that the product is safe and ethical for our listeners.

Q. Where do you think podcasting will be three years from now?

A. Podcasting will become a crucial part of any business trying to promote their products or services online. Technology is already making the ability to create podcasts easier. Three years from now producing a podcast will be an "on the fly" process that will be as simple as pressing Record.

Q. Did you have any formal training in broadcasting or radio?

A. Dave studied Radio Broadcasting in college, but that was a million years ago.

Q. How did you learn how to podcast?

A. Dave has always been a communications and new media geek, so it came easily. Community is also key, the Canadian Podcasting

Buffet hosts (Mark Blevis and Bob Goyetche) and listeners have always been available for advice and help.

Q. Why does podcasting appeal to you more than Vlogging, blogging, Message board posts or other social media on the Internet?

A. Podcasting is an actual recording of our voices. Our show is an audio baby book that our family will be able to hear for many generations long after we're gone. That's the main reason. We encourage our listeners to interact with one another by commenting on our blog. The interaction factor is of major importance to us; because it helps our listeners build community and get quick advice on many parenting matters.

Carrie Runnals, Suzanne Maiden, Sharon Martin, Amy Rissier and Julie Karneboge of TheDivaCast (TDC)

Q. How long have you been podcasting or broadcasting?

A. TheDivaCast hit the airwaves in April 2006. Suzanne and her husband, Robin (our producer, now dubbed "The Man"), podcasted Suzanne's Master's degree thesis (Insytworks) and introduced the rest of The Divas to the idea of an Internet talk show. We had absolutely no broadcasting experience, and no idea what we were getting into, but our desire to encourage other women to forge female friendships and find their "inner Diva" motivated us to jump off the cliff into the unknown realm of podcasting. Plus, at the time, it just sounded like fun. Little did we know…?

Q. What is your estimated audience size?

A. We look at the stats, but as you know, one of the challenges with podcasting is obtaining accurate download information. That is sure to change with the emergence of audience measurement standards, so for now, our ever-increasing subscribers, listener emails and voice mails assure us we're doing something right. Our target audience initially was the over-30 female crowd, but we've been pleasantly surprised at the amount of younger women and men who listen to our show. We are featured on the front page of iTunes's *Kids and Family* category and Podshow's *Family* directory. One obstacle we've encountered is many women within our target audience are not familiar with podcasting and are somewhat intimidated by technology. As this media becomes more mainstream and commonplace, we trust TheDivaCast will be more widely recognized.

Q. Tell us a little bit about your podcast.

A. We are five "everyday" real women with differing perspectives who get together and talk about real issues affecting women. We strive to encourage other women to find their own "inner Diva" and forge female friendships that Delight, Inspire, Validate, Affirm, and Support (DIVA). Through our Midlife Revelations (not Crisis) series, we urge other women to keep reaching and growing, no matter the age. All in all, we'd like to be seen as part of the collective sisterhood. We are an "infotainment" show, primarily focused on fun…a place where women can let down their hair and feel like they're hanging with their girlfriends.

Q. What motivates you to record your show?

A. We are motivated by our listeners, like I said, and our desire to encourage other women to find that "inner Diva" and forge supportive female friendships. Sometimes it is difficult juggling our busy schedules with five families—10 children—between us. We're all working mothers with endless to-do lists and at times getting together for fun let alone recording a show seems impossible. Podcasting has pressed us to reach further individually and even forced us to dig deeper, work through some group dynamics and ultimately become even closer friends. Producing TheDivaCast is sometimes challenging, but always worth it.

Q. How often do you tape/broadcast your show?
A. We're all working mothers, at times getting together for a show seems impossible, but we manage to put out a show weekly.

Q. Do you listen to other podcasts or mom shows?
A. Yes. We listen to both podcasting educational shows like Podcast Brothers, Managing the Gray, and Podcast 411. We enjoy Mom Talk Radio, Manic Mommies, Motherhood Uncensored, and Mommycast and we like to think of ourselves as older sisters to these popular mommy podcasters—we're the next generation.

Q. If you could describe one "emotional" feeling you receive from podcasting/broadcasting, what would it be?
A. A sense of deep satisfaction. Based on our listener emails and voice mails, we know we are touching thousands of women out there who relate to our "ordinary" lives. We try to convey openness and honesty with a touch of humor, so other women can relate to us and are encouraged to strive for deeper relationships with their girlfriends and families.

Q. Marketers are now looking for ways to get into the conversation of mom influencers like yourself and into podcasts. How do yo feel about that?
A. We're excited about advertisers' interest in podcasting. Even though the altruistic value of what we do is fulfilling, a little cash doesn't hurt to ease the pain of planning, scheduling, recording, and editing—producing a show is a lot of work. We've had sponsors in the past and we are aware of the related responsibility. Our listeners have grown to trust our opinions, so we feel strongly about not accepting potential sponsorship solely on money. We want to encourage sponsors who know our target audience and share our values and who can add to the lives of our listeners.

Q. Do you take paid advertising on your show?
A. Yes, TheDivaCast has been sponsored by some well known advertisers such as Splenda, Folgers, Marie Claire, and Leapster.

Q. What is in your opinion, the best way for a marketer to work with you or other mom podcasters?
A. LISTEN! As wives and mothers, we are the primary purchasers for our families—the CEOs of our households. Our listeners trust us to share truthful, reliable information about products. We think "conversational marketing" is most highly effective. With our past sponsors, we not only mentioned, "This show is sponsored by...," we talked about what we liked about the specific product within the conversation of the show. *THAT* is what helps sell products to listeners. Again, we would not make up this conversation. We would only allow sponsorship of companies that we honestly highly regard. That being said, if a company was interested in advertising with us and we'd never tried their product, they'd simply need to send us some samples. One good thing about TheDivaCast is we have FIVE women with varying opinions.

Q. Do you think you are jeopardizing anything by working with marketers to present new products, services, or samples to your audience?
A. NO! We only represent products/services we believe in. We try the products before we promote them, so we know they're reliable. We will not risk our reputation for short-term cash for a product that we wouldn't use ourselves. AND our listeners love give-aways.

Q. Where do you think podcasting will be three years from now?
A. Well, given the discussions about the recent hullabaloo about the term "podcasting," who knows? Regardless what it ends up being called, this new media is EXPLODING and we're excited to be a part of it, mainly because it gives "ordinary" people a voice.

Q. Did you have any formal training in broadcasting or radio?

A. No. We naively donned our Madonna-style microphones and started dishing about life. We've come a long way since those early days, and the sky's the limit for the future. That's the beauty of this medium.

Q. How did you learn how to podcast?

A. Suzanne's husband, Robin, is our producer and they recorded a psychology podcast, Insytworks, for Suzanne's graduate school thesis. Suzanne had the "intuitive hit" that we should take The Divas to the Internet airwaves, and TheDivaCast was born. At that time, Robin was the only one really informed about the medium, but since then, other Divas have initiated becoming informed about the industry.

Q. Why does podcasting appeal to you more than vlogging, blogging, message board posts, or other Internet social media?

A. Podcasting reveals The Divas' true personalities. Julie's perspective in her southern drawl, Amy's emotional plea to encourage donations for a fellow podcaster affected by house fire, and the infectious collective laughter of Suzanne, Sharon, and Carrie just wouldn't have the same effect without sound. The Divas can be spotted on various blogs and message boards of their individual interests. Social networking is an integral component of promoting podcasts. Some of the Divas, like Carrie and Suzanne, to date, have branched out to do some solo projects, like Suzanne's "Dear Zanny" advice podcast (sort of like Ann Landers of the Internet) and Carrie's Words-to-Mouth, a blog and complementary talk show dedicated to bringing "everyday women" readers and authors together beyond the printed page.

Q. What is the next media you want to venture into?

A. Video. We'll be adding video shows to our repertoire. Video will add a whole new dimension to TDC, especially for our Midlife

Revelation series where we push ourselves to try new things, like Harley rides, whitewater rafting, target shooting—activities that get us out of the Diva Den studio and help to remind us that at any age, no matter if we're moms and wives, we can still get together with our girlfriends and be a bit crazy and have fun!

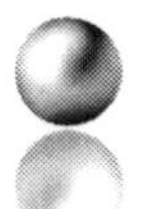

Megan Morrone from Jumping Monkeys: Parenting in the Digital Age

Q. How long have you been podcasting or broadcasting?
A. I started podcasting in April 2007. From 2000 to 2003 I worked as on-air talent on *The Screen Savers*, a cable television show on TechTV.

Q. What is your estimated audience size?
A. Around 10,000 per week, last time I checked. You can quote me on this, but I have no idea if it's accurate. I try not to be too concerned with numbers. As long as we have a dedicated community of listeners who give us feedback, I'm happy.

Q. How would your describe your podcast?
A. Jumping Monkeys is hosted by Megan Morrone, mother of three preschoolers and Leo Laporte, father of two teenagers. Each week we talk about the intersection of parenting and technology with helpful links, tools, and recommendations. We also feature interviews with other geek parents, contests, and more. We think we're funny. We hope you do too.

Q. What motivates you to record your show?
A. I love being a parent and I love technology. Talking about the way the two intersect is a good fit for me. Plus, I need a creative outlet. I also enjoy connecting with other like-minded parents.

Q. How often do you tape/broadcast your show?

A. Once a week.

Q. Do you listen to other podcasts or mom shows?
A. I listen to lots of other podcasts.The mom podcasts I listen to are Manic Mommies, Mommycast, The Parents Magazine Podcast, and Motherhood Uncensored.

Q. If you could describe one "emotional" feeling you receive from podcasting/broadcasting, what would it be?
A. Like any kind of performance, podcasting is exhilarating. My co-host Leo Laporte and I have worked together and been friends for years. Although we both have children, they're at radically different stages of their lives. I think this makes our conversations a little more interesting than the ones I have with parents of kids the same age. Also, our podcasts are pretty much unscripted. And Leo is known for saying whatever comes into his head. I never know what's going to happen. That's pretty exciting, to say the least.

Q. Marketers are now looking for ways to get into the conversation of mom influencers like yourself and into podcasts. How do you feel about that?
A. I have such mixed feelings about this. It's flattering to be approached by big companies and I'm willing to try products, but I try to make it clear that I'm not under any obligation to talk about what they send. If they don't want to send me things for that reason, that's totally okay with me. I come from a product review background, so a lot of this is not new to me. When I worked at TechTV we had people whose job it was to deal with companies and we could request products directly and we would return them when we were finished reviewing them. I have to say that I felt much more comfortable in that situation. Marketing directly to mom bloggers is such a new area. I really wonder if these kinds of marketing techniques will prove successful over the long term.

Q. Do you think you are jeopardizing anything by working with marketers to present new product, services or samples to your audience?

A. No, as long as the relationship with the marketer is disclosed. We have paid sponsorship on our podcast and it's clearly marked as paid sponsorship. Plus, we choose our sponsors and wouldn't have a relationship with anyone representing a product that we wouldn't use ourselves.

Q. Where do you think podcasting will be three years from now?

A. I wouldn't be doing this if I didn't see the audience for podcasts growing. I hope more people start listening and I think a lot of that depends on the language surrounding podcasts. Right now we say "subscribe," which on the Internet has traditionally meant that you're asked to pay for content. Also, many people access podcasts through the iTunes store, which also implies that there's a cost. I wish more people understood that podcasts are free, supported by ads, of course.

Q. Did you have any formal training in broadcasting or radio?

A. I worked as on-air talent on a cable TV show for three years. I wouldn't consider that "formal training," but it certainly prepared me for this experience. I feel like in 2000 we were one of the first programs to have a small, but extremely dedicated audience that we would connect with via e-mail, message boards, Web cams, blogs, etc. At the time our network didn't care about the dedication level of an audience, they were more concerned with numbers. I still care about numbers, but I think more and more big companies are recognizing the importance of a niche audience.

Q. Why does podcasting appeal to you more than Vlogging, blogging, Message board posts or other social media on the Internet?

A. Vlogging appeals to me, but it takes more resources than I have. I also think moms are more likely to listen to podcasts while they're at the gym, in the car, etc. It's harder to watch video during that time. I also blog, but I'm not really interested in social networking. With three kids, a husband, and a full-time job I have enough of an off-line social network.

Kelly McCausey from Work at Home Moms Talk Radio

Q. How long have you been podcasting or broadcasting?
A. I launched Work at Home Moms Talk Radio in November of 2003 as an internet radio show.

Q. What is your estimated audience size?
A. 1500 per week

Q. What motivates you to record your show?
A. Building a successful internet business with an emphasis on passive income through affiliate marketing helped me to pay off debts and replace my day job paycheck. As a single mom, being able to stay home full time with my son through his teen years has been great. The information shared on my show helps other moms achieve their goals to work at home and that is strong motivation for me. Every time another mom writes me to thank the show for inspiration and resources that gave them a boost, I know that doing the show is worthwhile.

Q. How often do you tape/broadcast your show?
A. Weekly

Q. If you could describe one "emotional" feeling you receive from podcasting/broadcasting, what would it be?
A. I gain a real feeling of connection from doing my show. My listeners are responsive. Since I do work at home full time, the

sense of community is a great advantage for me.

Q. Marketers are now looking for ways to get into the conversation of mom influencers like yourself and into podcasts. How do you feel about that?
A. My particular topic has always attracted a lot of advertisers and not all of them are acceptable. There are a lot of scammers who want to take advantage of moms who want to make money at home. I'm protective of my listeners and choosy about the ads we accept.

Q. What is in your opinion, the best way for a marketer to work with you or other mom podcasters?
A. Don't approach me as a marketer with something to promote asking for a spot as a guest on my show unless you have valuable content to share. "Infomercials" are not what my listeners are looking for. If you want to reach my listeners with your products or services, please support the good content of my show with your advertising dollars.

Q. Do you think you are jeopardizing anything by working with marketers to present new product, services or samples to your audience?
A. No. My audience is accustomed to my introducing new products and services that compliment the content.

Q. Where do you think podcasting will be 3 years from now?
A. Bigger and better than ever. More mp3 players, more media center PCs and more internet functions on gaming systems all add up to more available listeners.

Q. Did you have any formal training in broadcasting or radio?
A. None at all.

Q. How did you learn how to podcast?

A. The learning curve was pretty steep to start with. When I started it was internet radio. Podcasting came about a year later. I learned a lot by asking questions of people who were running hobby internet radio stations but most knowledge came by trial and error.

Q. Why does podcasting appeal to you more than Vlogging, blogging, Message board posts or other social media on the Internet?

A. I do blog and participate in all kinds of social media, but podcast is my favorite medium. I love to talk! Vlogging doesn't appeal to me; I'm not as comfortable in front of the camera.

Using Podcasting

Marketers who want to leverage podcasting can do so with little financial investment. It doesn't require an investment of equipment other than software, laptop and a microphone. There are a number of podcast platforms you may utilize to syndicate, measure downloads and manage the technical aspects of your show. One of the most popular is Libsyn.com. It provides wonderful measurement tools which allow you to see where moms are finding your show, as well as the number of downloads and total audience. Unlike traditional radio, podcasting allows you to more accurately measure the reach of your marketing efforts.

Once you have your technology in place, decide on a topic or show format. I suggest selecting one that is relevant to your brand. If you are a meal-planning product manufacturer, perhaps you focus on food and family. If you are a destination, you might consider a show on family travel. Keep it flowing with new and exciting information, guests and topics featured on each show. The show doesn't have to be an hour format. In fact, I would recommend a 20-30 minutes segment that allows time-starved moms to tune in for shorter amounts of time.

There are many ways to distribute or syndicate your show. You may rely on a site like libsyn.com to help you or you may upload new shows regularly to free sites such as iTunes. Make certain your show synopsis is compelling and interesting to moms because that summary is what sells the shows to moms in terms of time commitment. Remember that podcasting can be a global initiative so you will want to keep your comments broad in nature to appeal to a wider market of moms. Make sure you end each show with a call to action to visit your site for archived podcasts or additional information. It's all about integrated marketing so you want one marketing initiative to lead to another.

As Mom 3.0 continues to bridge together technology to gather information to use in her journey of parenting, shopping and entrepreneurship, podcasting will be a part of her tool box. She will utilize its mobility to take along relevant information and allow her to multi-task along the way. Companies who want to keep up with the speed of mom will want to consider podcasting and radio as means to tagging along with her. Customized content will grow in appeal to the Generation X and Generation Y moms who are customizing everything from products to advertising. Brand sponsored radio or podcast programming can give marketers a platform to speak to the individual needs of a mother in a unique and original format in a less crowded playing field. Podcasting and radio can be a valuable tool in connecting with Mom 3.0.

Today's Mom 3.0 is using video to share information, chronicle her parenting journey, learn from experts and peers and to simplify her life.

Venturing Into VCasts & Video

CHAPTER

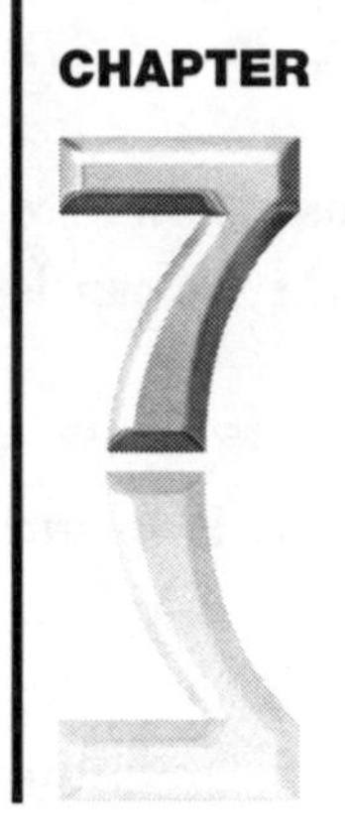

LONG BEFORE YOUTUBE OR WIKIPEDIA OR BLOGS, I WROTE ABOUT moms and information consumption. It was a decade ago. I predicted that at some point a mom would wake up in the middle of the night with a colicky baby and be able to go to her computer for a cure. Once online she would be able to select the originator of the solution, whether it is a physician, author or peer. The mother could then select the format in which she wanted to receive her medical advice. She could watch it, listen to it or read it. And if she happened to be in a park with the same colicky child, she could receive it via her mobile device.

This intuitive delivery of content is exactly what Web 3.0 promises us. However while developers are working on a flawless execution of this content delivery system, Mom 3.0 is piecing together technology to create this exact scenario. Video was the final piece of the equation and now moms are using the medium at a growing rate. According to BSM Media research, over 75% of moms have viewed at least one online video in the last seven days. In fact, BSM Media research points to online video as the next preferred media among moms. It's not a surprise if you think about it. Of course, moms are not the only ones watching videos online. Popularity of home-grown videos is growing daily. What's unique to the Mom Market is just how adaptable video is to a mom's life. Online video allows moms to hear and perhaps watch your message while stirring macaroni and cheese and answering homework questions! It's a perfect way to deliver your message in an entertaining, informative and easy to digest format. In this chapter we will discuss how moms are using video and how marketers can leverage this new soon-to-be mainstream media.

According to the Pew Internet and American Life Project, visits to video sites are up 45% from 2006 to 2007 and nearly half of all adults have used the Internet to view videos. eMarketer projects that by 2011, 86% of the US Internet population will consume online video, up from 62.8% in 2006. In raw numbers, this means the number of viewers will increase from 114 million in 2006 to 183 million in 2011. Interestingly, Latinos view more videos than African Americans or Caucasians.

Reaching beyond social networking, moms are embracing new technologies on the Internet and integrating Web 2.0 trends into their daily lives. In a recent study of moms nationwide, 80% of mothers are watching videos online, with 43% preferring videos that provide expert advice and solutions, suggesting that moms want to be able to find what they want when they want it. Video among moms is popular, not only because of the convenience and quickness of the content delivery, but it also allows moms to share and archive her parenting journey.

The five core values of moms are again motivating the behaviors of mothers. In fact, a point I should have made earlier—the more values a medium or marketing program hits upon, the more effective it will be in motivating a mom to use it, or share it with others. Before we discuss how you can leverage video to win over mothers, let's examine how moms are using video. I think what you will find is that your consumer is your best teacher.

Uses of Video

Video has always been a medium loved by parents. It brought our memories to life, but it has never been particularly easy to use. If you are a Baby Boomer, you probably remember your own parents taking out the large family movie machine and watching your dad thread the film through the wheels. Quality was iffy at best. The screen was cumbersome to set up and family movie night often took hours of preparation. I remember my own parents giving up on the screen and resorting to a white sheet wedged into the back of door. Heaven forbid if someone had to use the bathroom in the middle of film—we'd lose the screen and another 30 minutes trying to get it up again! In the 90's, video cameras became the fad but required an adapter to play tapes or cords to the television. Few of us really took the time to plug in the red, white and yellow wires into the TV so we ended up with boxes of unwatched tapes. For parents who currently have teenagers, they must now convert all the handheld video tapes to DVDs in order to archive their memories—a cumbersome task at best. For most families, a stack of old beta tapes sit in a box at the top of the closet. Fast forward to today and I'm sure you'll agree a mom's use of video has certainly changed.

Today's Mom 3.0 is using video to share information, chronicle her parenting journey, learn from experts and peers and to simplify her life. Moms are even firing up their handheld camcorders and creating marketing videos for their business. In desperate times, they are using their webcams to entertain their children and then emailing files to grandparents on another coast. With Flip cameras in tow, they are taping product reviews in the aisles of superstores, parks and school yards. Many bloggers are now becoming Vloggers while mom podcasters are stepping in front of webcams to become Vcasters. They are laughing at YouTube videos and catching up on missed episodes of LOST on ABC.com. I even know one mom who uses her webcam to recite homework assignments with her children. When it's time to search for a stroller, they are watching company produced product videos and then taking in a few mom produced product reviews.

Moms are also using video to build businesses. About a hundred or more moms are now Vcasting and hosting their own online TV shows on a regular basis. Although the numbers are currently small, the trend is catching on and I foresee a day when there are literally hundreds of branded online mom shows. The ease of Flip cameras has enabled moms to do more with video. Mom bloggers are now adding videos to their blogs by taping product demonstrations or customer experiences. Small business owners are filming promotion clips and distributing them via widgets. They are also using video clips to promote themselves on Facebook, LinkedIn and MySpace. Home-based working moms are using video to keep in touch with virtual employees and to pitch new clients. Needless to say, Mom 3.0 is creative and she is using video in every possible form of the technology she can think of.

I think the most exciting new behavior on the horizon is family video archiving. There is a new wave of video sharing sites for moms emerging. These sites leverage the strong need that mothers have to archive their journeys through motherhood. As the Chief Family Archiver, moms are constantly searching for easy-to-use solutions for digital photos and video. The emotional tie to digital imaging is so great among moms that I've devoted an entire section to it later in the book. Best of all, you don't have to be a photo-related company to leverage this strong motivator to sell product or connect with mothers.

Video Sharing

One example of a video sharing site is SeeToo.com. It's a free service that allows moms to share videos with family and friends across the US and then simultaneously watch them together. It combines the appeal of blogging and allows viewers to comment about the video in real time. The only flaw of the site right now is that once the viewing session is over, the participants in the share group can no longer see the video. The site also allows families to share photos and music. In the Mom 3.0 world, a site like this would suggest other like-minded viewing groups to the user so that she can share her knowledge, as well as suggestion other videos that she herself might like to view. For instance, SeeToo.com would be intuitive to know that if the host mom is sharing video of her family's Disney World vacation, she might enjoy sharing it with moms looking for Disney World vacation planning information. They would also offer product suggestions based on the content of the consumer-generated content and allow the viewers to compare purchases. In the Disney World example, SeeToo.com might offer photo album books or other post vacation items. This would be truly intuitive to the mom using the service. It would also allow her to archive her child's video alongside electronic photo albums and scrapbooks so that a memory could be relived in multiple forms. The site would basically multi-task with the multi-tasking mom.

I have many bad habits ranging from Oreos to Diet Coke consumption, but when it comes to Marketing with Moms my sin lies in impatience. If I see an untapped opportunity that can be supported with the research, behavioral insights and consumer trends I possess and I can't convince anyone else to tap it, I normally go out and do it myself. In the case of the emergence of video and the viability of it in the Mom Market, I built Newbaby.com. Unlike most of my endeavors, this one was too large to take on alone and it was with the help of my business partner, Bob Sullivan, which brought the opportunity to life. It will serve as a good example of almost every way a mom is using video rolled up into one site. I invite you to spend some time online clicking around. Register to gain the full appreciation for the sharing, archiving and video community it provides mothers.

We launched NewBaby.com on Mother's Day of 2008. In the shortest terms, it is an online video community where moms learn, share, connect and create with videos. It offers every type of video from expert advice to peer product reviews to corporate product promotion. We have moms keeping public video diaries, chronicling daily their journey through motherhood. One Vlogger, as we call them, has been taping her life every day since the birth of her baby over a year ago. Each and every day, viewers get an update on what's going on in her family's life. It has the features of a social network, but enhanced with video capabilities. Moms can register and create a profile page. Once registered, she can browse the video library, upload videos from other sources or record videos. She can even save a video on her profile page to view later after the kids go to bed. My favorite feature is the Family Video Gallery. It is here that moms can store videos, choose whether they become public or private, and share select videos with friends and family. Best of all, it's safe, secure and free. In addition to the video engagement on Newbaby.com, moms can also access podcasts, expert blogs, RSS feeds and connect with other moms.

Remember my earlier description of an online destination where the mother of the colicky baby could choose the delivery of her content? We have created this on Newbaby.com. Should a mother search colic, she will be offered video, articles, podcasts and blog posts authored or produced by experts, corporations or other moms.

As a Mom Marketer, I took into consideration the needs of companies trying to connect with moms and there are several opportunities built into the site. First, companies can upload product videos free of charge. Consider it YouTube meets Moms.

It's a great way to promote your brand or demo a product. Secondly, there are the more traditional means of advertising such as online polls, sponsorships, banners and buttons.

The most revolutionary opportunity can be viewed in the videos produced by NewBabyTV. Animated ads with customer lead generation software built into them are laid over the video to, in a non-intrusive manner, speak to the viewer. Finally, marketers can create customized widgets to place anywhere on the Internet and draw video from the Newbaby.com library. We will talk more later about leveraging these and other video opportunities for you company. First let's hear from some of today's most popular Vcasters and online TV hosts.

Most Popular VCasters

Kimberley Clayton Blaine, MA, MFT
Founder and host of TheGoToMom.TV

Q. How long have you been podcasting or broadcasting?
A.10 Months

Q. What is your estimated audience size?
A. 5,000 viewers per month

Q. What motivates you to record your show?
A. I feel that American parents are entitled to free expert advice on the best parenting practices and what better way to deliver than by podcasting my Blogs, Vlogs, Video segments and radio show.

Q. How often do you tape/broadcast your show?
A. Video tape twice per month and radio show once per month.

Q. If you could describe one "emotional" feeling you receive from podcasting/broadcasting, what would it be?

A. Exhilaration

Q. Marketers are now looking for ways to get into the conversation of mom influencers, like you, and into podcasts. How do you feel about that?

A. I'm so excited and really would love to work more closely with corporate America to enlist them in the important task of helping educate parents about our children's well-being.

Q. Do you take paid advertising on your show? If so, what kind of packages do you offer?

A. On my main site (www.thegotomom.com) I do not allow advertising. I license my show to other large dot coms, and generate revenue that way, since they already have marketing/advertising relationships in place. I let them do the work and do a contract plus revenue share.

Q. What is, in your opinion, the best way for a marketer to work with you or other mom podcasters?

A. Marketer should know exactly what we do and stand for. And pursue the relationship year round so that both benefit. Collaboration both ways. There's nothing more disappointing then a marketer not knowing what our brand stands for or what we do. Research is key here. We are all so busy, so we need to hit the bull's-eye each time.

Q. Do you think you are jeopardizing anything by working with marketers to present new product, services or samples to your audience?

A. If I pushed a product that I clearly did not believe in, I'd feel that I sold out or may have jeopardized my reputation. However, when I am matched up with good products that I believe in, I truly feel I am helping the economy, myself and American parents.

Q. Where do you think podcasting will be three years from now?

A. I feel that it will be the standard for anyone who has the platform. The question is, who are the professional podcasters vs. the novice podcasters?

Q. Did you have any formal training in broadcasting?
A. None. I am a child therapist.

Q. How did you learn how to video cast?
A. I trained myself by listening and watching others. Since my show is video based I had to research how to deliver content in a quick and sufficient way to keep my viewers engaged.

Tracey Henry, Suburban Diva, Vcaster on Newbaby.com and Suburban Diva.com

Q. Describe your video/online TV involvement.
A. I do video blogs and video product reviews for Newbaby.com.

Q. Did you have any prior television or on-air experience?
A. I have no on-camera experience. I keep a written blog, as well as write a weekly syndicated humor column. I've also written a book based on those columns, but I tend to be a little camera shy.

Q. How often do you tape segments?
A. I try to put something up at least once a week.

Q. Can you describe the production process?
A. A video blog is simply me sitting down to my computer in front of a webcam. It's a place to record my current thoughts and experiences as a mother, much like a written blog. I have a vague idea of what I want to say, but I don't write it all out, I just try to speak to the camera. Product reviews are a little more technical, and actually include a camera person. These are done in my home with some friends I've recruited for the process. I will generally

write out a little more ahead of time, but there is nothing scripted. We simply speak to each other honestly about the product, and maybe give a demonstration.

Q. Why do you believe video is a good tool for connecting with moms?

A. This is a complicated question, but it really revolves around gaining a new level of intimacy with an audience. With video, there is a level of authenticity that can't be duplicated in other communication. To see that there is a real person, a real mom just like you, using the product or going through the same experience is quite powerful.

Q. What opportunities exist for marketers to leverage video to connect with moms?

A. In addition to participating on a site dedicated to giving new moms the latest information, I think the product reviews are the perfect outlet to get products directly into the hands of their target audience. There are also other innovative marketing tools on the site including video "walkovers" which are effective. And of course there are the traditional banner ads available as well.

Q. Where do you think video will do in the future?

A. I think it will only become a stronger, more powerful and sophisticated marketing tool. The potential of this medium is limitless, and really offers something especially effective to advertisers.

Kris Jackson
Co-Founder, Clever Parents TV

Q. Describe your video/online TV involvement

A. Clever Parents TV is the video complement to Clever Parents (www.cleverparents.com), a website for parents that my husband,

David, and I founded in 2005. Clever Parents TV episodes are the "best of" CleverParents.com. David and I sit at a table facing the camera and, in a conversational style, talk about what is current on the website. We also include interviews with parent entrepreneurs and some of our columnist experts.

Q. Why did you decide to do Clever Parents TV with your husband?
A. Because it's fun! David and I have a blast shooting the Clever Parents TV segments and we are generally trying our hardest not to burst out laughing.

Q. Did either of you have any prior television or on-air experience?
A. No.

Q. How often do you tape segments?
A. Prior to the birth of our third child last fall, we taped weekly. Having three children under the age of four slowed us down, but we plan to get back to weekly tapings soon.

Q. Can you describe the production process?
A. Bare bones! We shoot in our basement with a camera, remote, lavaliere microphones, and a lot of lights. We are not scripted and we typically go with the first shoot unless one of us completely flubs something or starts cracking up.

Q. Why do you believe video is a good tool for connecting with moms?
A. I think video is a good tool for moms for two reasons. First, video is personal and it's accessible. Secondly, it's good for time-starved moms. Parents are generally strapped for time and trying to accomplish as much as possible in short increments. Having video available online makes it possible to get the information, or entertainment, they need quickly.

Q. What opportunities exist for marketers to leverage Clever Parents

TV to connect with moms?

A. On the Clever Parents and Clever Parents TV websites, we only write and talk about products that we would have in our own home for our families. For that reason our viewers trust us. We receive a lot of products for potential review and we are always interested in hearing about new products, services and partnership opportunities.

Q. Where do you think video will go in the future?

A. I think that the use of online video will become more and more pervasive across the Internet. The tools are getting easier to use—Mac users are fortunate that computers come loaded with an iMovie package that makes it really easy to produce quality work. I think we are at the very beginning stages of online video and we'll look back at this time and be surprised at what a difference a few years make.

Opportunities with Video

Now that we've heard how moms are using videos, let's explore the opportunities for you to utilize this new media to connect with Mom 3.0. Basically your options fall into two categories: self produced video or consumer generated video. Regardless of who is the lead producer of your video strategy, the content should accomplish one of the following: entertain, educate or engage moms in sharing.

BSM Media research shows that 75% of all moms watch videos for answers. They like to view solution-driven videos that are no more than three minutes long. Occasionally, their appetite for information will keep them watching for more than three minutes but no longer than five minutes. Even though online video is relatively new, consumers have already been trained that online segments should be condensed. In fact, what Bob and I have seen on Newbaby.com is that moms will click on as many as 11-15 different videos in one viewing session, one right after the other. Of course, the more entertaining the video, the longer they will watch.

One of my favorite product videos is produced by Bugaboo International stroller company. It was carefully choreographed to show off the durability of the stroller while good-looking 30-something dads dance through the streets of New York City with the stroller in tow. It's really entertaining and amazing to watch. I am not in the market for a stroller but I can assure you I've directed no fewer than a dozen friends who are to the video. Not to mention I've just told each of my readers about it. To view it yourself, click over to Newbaby.com and search for "Bugaboo."

Engaging moms with your video goes beyond just having her recommend it to friends or view attentively. The pages of trade publications are filled with examples of marketers who have engaged consumers by challenging them to create branded video for them. Dove's Campaign for Real Beauty campaign did it when they asked women to submit commercials for the 2008 Super Bowl. Doritos also successfully engaged consumers by creating advertising spots. The outcome not only earned Doritos free creative for a Super Bowl ad but also generated a lot of buzz as consumers viewed popular entries.

Producing Your Video

The challenge for companies who want to leverage video is the cost of production. While moms have no problem viewing homemade videos from mom Vcasters or Vloggers, most companies would not be comfortable putting up anything less than professionally produced video. I would challenge this manner of thinking. Perhaps the best example of home video becoming great marketing is the infamous Mentos-Diet Coke videos that appeared on YouTube. They were simply produced, but received millions of views from potential consumers. Don't over think your videos. Yes, they should be interesting. Yes, they should be relevant and useful. Yes, they should be entertaining. However they don't always have to involve spending thousands of dollars or scripting every word of the content.

Think about the popularity of *Sex and the City*. For male readers, you may not understand the draw to Carrie and the girls but here's my best explanation. The appeal of *Sex and the City* is that conversations are light and uninhibited. Female viewers feel as they could be sitting at the table, trashing guys and dishing about fashion as easily with Sarah Jessica Parker as with their best friend. This doesn't mean you should go out and act like a best friend to every potential mom consumer but it does mean that casual is okay. It's real and real is what appeals to today's Gen X mothers.

I know larger companies are going to run out and use their agency to make videos, but there are options for smaller brands without big budgets. The first option allows you to use professional production staff in the off-hours to produce great quality videos. I've never met a producer or cameraman who wasn't willing to freelance. Check Craigslist to find a freelance cameraman. In most cities they run anywhere from $75 to $200 per hour. The more expensive part of video is the actual production and editing. Here too you can find a freelancer or a small scale production company. I find that the going rate for video segments is about $1000-2000 per 3 to 5 minute web-ready segments.

Video Advertising

It was reported by *Online Video Advertising* that advertisers will spend at least $1.5 billion on online video ads by the end of the decade. The challenge for marketers, however, will be to create, customize and adjust the creative for each video platform online without substantial expense. This means that any change in the direction of a campaign requires expensive production and multiple teams working in tandem. It does, however, make marketers more honest to their goals. No longer can an advertising team create a print ad and blanket it into various books.

Video platforms tend to be even more specific to the lifestyles of its consumers, requiring marketers to customize. There are several ways to maximize advertising opportunities with mom videos. First, you can simply attach a branded trailer to the front or the back of the video. A good combination of resources is to produce videos with a mom spokesperson, but introduce it with a company trailer. So for instance, if you are a baby furniture company, you might want to enlist the help of someone like Elizabeth Pantley, who is a mom author of several books about sleep. You can produce a dozen solution-oriented videos for moms on topics from "getting baby to nap" to "sleeping through the night" and have each one begin with a message from your company.

The other video advertising option is more like traditional commercials that play before the video actually loads. This method of advertising is most common on sites like Yahoo!, iVillage and AOL. A non video option is to purchase ad space such as banners and buttons on the video viewer or on the page which houses it.

The next generation of video advertising can actually be seen on Newbaby.com. It is an interactive in-video lead generation system which allows the viewer to click on a product-related flash icon and almost instantaneously purchase the product. I invite you to see it for yourself at www.newbaby.com. The interactive video credit card transaction campaign enables the users to transact within the experience without the user leaving the video. This function complements video offerings featuring product placement focused on the advertiser's goods and services. The in-video direct option of advertising allows the viewer to purchase product within a secure environment without taking her outside of the video frame.

For advertisers, in-video options allow you to place your product in front of moms who are highly qualified and extremely engaged in subject matter related to your product. It is cost effective and unobtrusive and the customizable option allows advertisers to reach specifically targeted moms with information and offers within contextually relevant video segments. Interested users opt in through interactive elements embedded in the video player, momentarily pausing the video while allowing the user to engage the informational offers or surveys within the video player. When the user completes the action the video resumes, leaving the consumer with the information and opportunity they were seeking in one interaction. With a Supervised Contextualization® service you can ensure every video receives the correct meta tags on the video timeline, regardless of how deceptive the sound track is at matching offers to content. Each video is automatically pre-processed then individually screened to deliver the congruent advertisement that will give the maximum CPM on your video placement. It's a very forward-thinking way to sell to consumers.

Let's use the example of the baby furniture company we discussed earlier. The baby furniture company has several videos on sleep uploaded to Newbaby.com or other video sites. With the direct in-video advertising option, the mom will see a small non-intrusive screen pop up into the video. In some cases it may be an animated "Avatar" (a customized character) walking along the bottom of the screen holding up a sign for sleep monitors. The mom who is watching a baby sleep video sees the in-video ad and clicks on it to purchase the item without leaving the video viewer. In less than three minutes, she has learned how to get her baby to sleep through the night and purchased a monitor for her baby's nursery. It's very similar to the next generation of direct interactive television applied to Internet video.

Widgets and Distribution

Once you have videos produced, widgets are a great way to distribute them to moms. According to the New York research firm eMarketer, marketers spent $15 million creating, promoting and distributing widgets in 2007. That number is expected to exceed $40 million in 2008.[1] Widgets are mini-Web applications that are installed or downloaded by the user. They can be installed on Web pages, blogs, personal computers and social networks pages like Facebook or TheMotherhood.com. Once the widget is installed, it becomes a delivery tool for updated information that can be aggregated from several sources; thus eliminating the need to visit multiple sites. Moms can install a widget and have weather, news and blog updates fed into it, saving a great deal of time.

Since so many widgets are used within social networks, they present an effective marketing tool to reach the highly active social community of online moms. The opportunity in this category is to create a widget that is branded and delivers relevant and useful information to mothers. Often the content is created around a common theme. For instance, Hewlett-Packard has a widget that allows moms and others share picture-taking practices with each other. As content is updated, it is sent out through a widget to the consumer who develops a relationship with HP. I think the opportunity in the Mom Market also includes filling a void for video content. In the late 90's, when webmasters were trying to produce enough information to fill websites, content was king. I believe that the same sense of urgency exists today with video content on the Web. However, this time around the content is expensive to produce so if you have the content, there are plenty of opportunities to distribute it. You could essentially fill the needs of mom site owners who can't afford to produce it themselves by offering it up through a widget. It might mean that you have to think of yourself in terms of a content syndicator, but the result could mean getting your brand in front of millions of moms.

Newbaby.com was built to allow moms, as well as companies, to create free customizable widgets. The widget is a portable video

---

1 "Data Points," *Brandweek*, (February 4th, 2008), 8.

player that allows moms, bloggers and companies to show their favorite NewBaby videos anywhere they want. Moms can even give their child's grandparent a widget to watch homemade videos of her grandchild. Once a mom has a widget, she can create a playlist with videos she's uploaded to her profile on Newbaby.com, choose her own colors or logo for the widget and install it on her blog or desktop. Additionally, each time she adds a video to her profile, it automatically populates her playlist. Mom 3.0 can now record her baby's first words with her Flip or webcam and share it with family and friends around the globe in a customized viewer by way of her widget.

Another method of distribution is through search optimization. The challenge with search and videos is that the functionality of search has not yet caught up with video. Unlike an article, which allows a spider to log related words, they are unable to evaluate audio content. To gain the best placement of your videos, marketers must assign the most relevant meta tags. The description of your video is extremely important to gain proper placement on video sites such as Yahoo! video. Newly developed video search capabilities on the Internet are quickly becoming a preferred means to obtain relevant information for today's Mom 3.0. From startup technologies such as Blinkx and Pixsy to search giants like Google, AOL, Yahoo and MSN, video search is one of the fastest growing percentage segments within search and advertising.

Companies Using Video

Let's take a look at some other companies successfully using videos to connect with moms. Johnson & Johnson recently decided to enter the mom video arena with the launch of www.touchingbond.com. Surprisingly, this conservative, veteran mom brand took a new and hip approach by using Web cartoons. In one of the new animated videos a mother massages the feet of her baby, which evokes laughter, giggles and smiles from her daughter. The cartoons successfully play up the connection between mom and baby. It is a clear switch from the solution directed messaging J & J has used in the past to connect with moms. The videos entertain, engage and educate all at once.

As mentioned previously, Suave and Sprint partnered together to take another approach by creating In the Motherhood (InTheMotherhood.com). Visitors to the site can view an online video series starring celebrity hosts—Jenny McCarthy, Leah Remini, and Chelsea Handler—who inhabit the mother "hood." Suave engages moms by conducting an episode-writing contest each year, with the winning storyline becoming a future episode. Additionally, moms can post comments about entries, current episodes and topics relevant to mothers who also are visiting the "hood."

Finally, the folks at Meredith, the publisher of *Better Homes and Gardens*, *Family Circle* and *Parents* magazines transformed their content into video and aggregated it at Better.tv (www.better.tv). For moms there is a series called MiniMomCast which utilizes some of the print experts from Meredith's collection of family titles.

Tomorrows Next New Media, Today

Like a good marketer, I take my cues from my target consumer group. I watch their behaviors to determine how I should speak to them and how to deliver my messages. Mom 3.0 is on the road to video and I encourage marketers to follow her. It's not too late to catch up. Video is young enough as a technology that if you adapt it now, you will likely be creating resources that can become part of mom's videocast behaviors while they are still in the development stage.

Today I am watching hundreds of mom bloggers adding video to their posts each day, with the help of Flip cameras and webcams. Moms who otherwise live private lives are taping segments and uploading them to sharing sites. More importantly for marketers, moms are turning more and more to video for product recommendations and everyday solutions. As the supply of mom video is still low and demand high, there is definitely an opportunity to be the first to market. For companies who want to be forward-thinking with today's Mom 3.0, video offers an excellent opportunity to connect with mothers.

The opportunities in mobile communication are numerous and will continue to grow as functionality increases in wireless devices.

Moving Into Wireless With MobiMoms

CHAPTER

THERE'S NO OTHER TOOL MORE IMPORTANT TO BRIDGING TECHNOLOGY for Mom 3.0 than her wireless device. Her mobile phone, whether a Razor or Blackberry, allows her to integrate her online behaviors with her on-the-go lifestyle. Moms are using their wireless devices to manage busy schedules, prepare dinner, facilitate business deals and stay in touch through micro-blogging. They are sharing photos, capturing videos and networking with friends. We have already established that moms make up a great part of the social networking population in the US. This means that the Mom Market is primed for mobile and wireless networking. According to a study by Ipsos[1] titled, "*PROFILE: Social Networker*," US social networkers are significantly more likely to own other technology, consume more digital entertainment and own mobile devices. Over half of US social networkers have used their mobile device to send or receive SMS text messages or emails. They have also used their devices to browse the

1 Ipsos Study, *PROFILE: Social Networker*, www.marketingcharts.com, (accessed December 6th, 2007.)

Internet and receive digital images. Ironically, even though internationally mobile devices are more feature-rich than in the US, users in the states leverage their devices more to communicate with family and friends and manage life's tasks and events. Not only do they use their devices more, they also have more of them when you account for DVD players, digital cameras and game systems. The wireless device is Mom 3.0's communication center.

Opportunities for New Marketing and Moms

Mobile marketing with moms is what's next on the horizon. And I will take the liberty, with the help of a few experts, to extend some forward-thinking ideas of where mobile marketing with moms will ultimately emerge. There is risk, however, in predicting where new marketing is going because in the time it takes to publish the words we type today, many of these tactics will have become reality. Notice that I've used the word, "will" rather than "might." Consumers are driving marketers to change and they are doing it at warp speed. Now that moms are familiar with technology and its capabilities, they expect companies to utilize these tools to become more integrated into their lives. Complete wireless functionality is probably the most anticipated technology among moms. They know that if their wireless device eventually suggests dinner ideas, interacts with commerce as a payment system for her children and provides coupons for product in the grocery line, that her life will be that much more efficient.

Wireless is the richest area of opportunity in the future. However, it will require wireless companies and their potential partners to truly understand the life of a mother in order to cash in. As long as they view their devices solely as a communication tool, rather than a solution ecosystem, they will miss developing deeper relationship with moms. The wireless industry will simply remain a promotion and price-based game. Today's mobile devices with tomorrow's available technology are perfectly aligned to meet the needs of a fast moving Mom Market—today and tomorrow.

Consider these ideas from the prospective of a mother. Take meal planning. Eighty-two percent of mothers do not know what they are cooking for dinner at 4:00 in the afternoon. What if there was a way for moms to order curbside service via a text menu from her cell phone? Dinner could be picked up on the way home and the family meal planning dilemma would be solved. To take the idea one step further, what if each family member could text the restaurant with their menu selection and the restaurant could merge the family's order into one? This would save mom time, help her feed her family and develop a strong loyalty to the restaurant's service or product.

If the wireless company decided to base their services on life stages of moms, they might provide her with e-birth announcements. This would allow a mom to share the news of her new arrival without worrying if dad made all the obligatory calls. Through customization of services and the ability to personalize the experience with the wireless device, companies will be able to form deep, meaningful relationships with moms. Wireless companies will tell you that functionalities such as these are months, perhaps years away. There are, however, many wireless applications that exist today, but are not being leveraged to maximize marketing efforts with moms.

The most immediate opportunity exists in the photo function of a mom's cell phone. Although the camera feature is almost standard in today's mobile phones, our research points out that very few moms are actually using their phones to capture images that they later print. However, younger Generation Y moms do use the photo sharing function. It's not uncommon to see Mom 3.0 snap a cell phone photo of a dress and send it to friends to gain their opinion before purchasing it. What's not happening easily, and presents the greatest opportunity to a forward thinking company, is creating a click to print application that is easy and simple to use. Most moms would agree that today's cell phone camera is like the VCR clock of the 1990's; everyone has one but no one really knows how to set the time on it. Our research shows that very few moms really know how to take the photos off of their cell phone. Other than capturing a screensaver or feeling less guilty for having forgotten the camera at the school play, the camera functionality falls short.

Another opportunity in the world of wireless is mom scheduling solutions. The number one reason after safety for moms caving on the "mom can I have a cell phone?" debate with her children is managing the family's busy schedules. She has a need to stay in contact with all the members of her family and empowering each member with their own mobile device helps her do this. Today's Mom 3.0 is bridging together communication by supplying family members with free email accounts, access to texting and, of course, cellular devices. For marketers in the youth cell phone market, BSM Media research indicates that one of the barriers to purchase is the fear of uncontrollable cell phone bills. To speak to this issue is to overcome a key objective.

The opportunity for marketers is to create a platform that unites her offline schedule with an organizational system that transcends PC and wireless functionality. There are some companies who are pursuing this, but what mom really needs is an electronic platform that allows her and her family to merge every individual schedule into one. The platform would additionally allow her to combine and share photos, videos and text content.

The next generation of wireless will deliver moms content in the form she wants it when she needs it most. For instance, a mother is at the park and her child is stung by a bee. She can take out her wireless device and search content which delivers a short solution based video because she doesn't have the time to read a lengthy article. Later, she is at the grocery store and needs a recipe for ham casserole. Her wireless device not only delivers a text recipe complete with a shopping list, but also an online coupon for the noodles. It's intuitive and delivers information in the form she needs it in the form she wants it.

Texting

Texting is a major part of a mom's tool kit in the wireless world. Boomer moms most often adapt to texting when their tweens or teens introduces them to the convenience factor. Younger Generation X and Millennial moms have been sending text messages since they were teens and are quite adept with their fingers. According to a 2007 survey by M: Metrics, Hispanics are more likely to use mobile phones for purposes beyond making phone calls. Over 6% of Latinos use their mobile devices to view videos while only 3% of the general public does the same. When it comes to sending photos, 26.8% of Latinos send photos or videos while 15% of the Anglo population takes part in this activity.[2]

Companies should not overlook the opportunity to have moms text responses to surveys and products. We've all seen it work with interactive text programs, such as *American Idol* and *Deal or No Deal.* There are many ways to establish a dialogue with moms through texting interaction. For instance, a mom blogger can ask her audience to vote on the next topic of discussion by texting their topic choice. A company who is developing a new advertising campaign can show the candidates on their website and then ask moms to vote via text message. For the mom on the go, it's a great way to interact with her as she's moving through her day and utilizing the technologies that she takes along with her.

The Future of Wireless Marketing

Before we leave the topic of mobile devices, I want to turn to a woman who has impressed me more than any individual I've met in the wireless world. She truly is a forward thinker who not only hypothesizes about what mobile marketing can be in the future, but actually implements on the existing functionalities. Meet sophia stuart, who spells her name in lower case letters intentionally. She is

2 *Ad Week Media*, April 7, 2008 Special Report, Page 1, "Hispanic Marketing Report"

the Director of Mobile Marketing for Hearst digital media, Hearst Magazines. I recently had a chance to sit down with sophia to speak about the future of wireless marketing and I'd like to share her insights with you now.

sophia stuart, Director of Mobile Marketing for Hearst digital media, Hearst Magazines

Q. Marketers keep reading about Mobile Marketing, yet it seems undefined and slow in coming. Where is it and when do you think it will finally arrive?

A. In a recent Pew Internet and American Life report (March 2008), consumers said their mobile would be the hardest tool to give up rating harder than Internet or TV. Mobile is not *yet* a daily tool for consumers, but it will be and that's when the mobile marketing dollars will really flow. The best analogy is that we are in the same stage for mobile as we were in the mid-90's for the Internet. In 1995 it was still mainly men (and splendidly forward-thinking early adopter women) looking at news, sport, weather and technology sites

If you look at the information that people are consuming on the mobile web today, they are many of the same content areas that dominated Internet searches in the 90's. But remember 2001, when women became the dominant audience on the web? That holiday season of 2001, women tipped the balance and became the majority of shoppers using the web and then it took off. So we need a few things to happen in mobile before it becomes ubiquitous: flat-rate pricing for all-you-can-eat mobile data plans (unlimited text messaging and mobile web surfing and decent mobile video) as well as mobile broadband (fast Internet-type speeds to your mobile phone to make surfing a real pleasure). Then we will see marketing take off on mobile devices as a result of consumers using their mobile as a daily tool. After all, it's the only media you carry with you 24 hours a day.

Q. What do marketers need to know about Mobile Marketing?

A. The most important way for marketers to understand mobile marketing is to use a range of mobile devices (not just high-end ones!) and really start to explore the on-the-go experience and then interview every female consumer they can find to see how they are using their mobile device. When we developed the mobile sites at Hearst, we did ethnographic work to identify unmet needs. The question we asked ourselves is what do our various audiences of women need on the go: from teen girls (*CosmoGIRL!*, *Seventeen*) to young women (*Cosmopolitan*, *Marie Claire*) to women over 35 (*Redbook*, *Good Housekeeping*, *House Beautiful*, *Harper's Bazaar*). That information shaped how we created our special sites that are purely made for mobile use. Mobile web is not about taking Internet content and squishing it onto a tiny screen. That is a truly dreadful experience. The content needs to fit the technology format.

Women tend to use mobile devices as a way of creating personal space. The amount of women faking a phone call to avoid talking to someone approaching is higher than you think! This also influenced our Cosmo Fake Calls service where a young woman can book a call from us to get out of a meeting or a date. Their mobile rings and it is entirely scripted to get her out of whatever sticky situation she is in.

Moms also use mobile devices as a means of companionship, which is why we have terribly popular blogs on all of our mobile sites at Hearst. Women love to spend a few minutes catching up with an unfolding storyline so blogs like our "Infertility Diaries" or "Diary of He" on redbookmag.com are very engaging.

Marketers should also know that once a woman is equipped with information at her fingertips, she is going to use it. I just got back from a trip to Lisbon and saw a number of mothers doing searches on the green credentials of certain baby products while standing at the aisles in a supermarket. They then made a purchase choice on the baby products that were the most "green."

Women have not had access to information like this ever before. So we developed robust product guides on our mobile sites, especially from the Good Housekeeping Research Institute, so we could give them the results from our consumer product and services testing. Ten best coffee-makers, ten best querty keyboard smart phones and so on, all tested in our GHRI labs in New York.

I know of several women who have changed their airline choice because another carrier had launched their mobile sites with check-in, upgrades and information. If a marketer is not there on the mobile web, the female consumer is going to switch her brand to someone who wants to make her life easier.

Q. For companies who want to be ready for the opportunities that Mobile Marketing will offer, what should they be doing now?

A. We recommend that they start building mobile sites that are made-for-mobile which means simple to use, packed with information that a consumer needs wherever she is and has a feedback loop so she can tell you what she thinks about your product, get answers to FAQs and make good decisions. Quizzes are a great way for mobile consumers to identify which product in your range is for them. For example, if a marketer is looking to showcase their skincare regime, have a mobile site with a quiz that lets women input which skin type they have, what climate they live in and so on, so you can make a "personal recommendation." Then why not link up with a digital retailer so she can make the purchase right there from her phone? Or give her the information she needs to find a store nearby that stocks the recommended product.

We did some SMS quizzes for a Hollywood studio that wanted to promote their on-sale date of two major DVDs. We got a very high response rate from users who opted-in to play the quiz, which was all about engaging them with the themes of the movie so they would want to purchase the DVD.

Q. How do you believe Mobile Marketing will change marketing as we know it?

A. We have never given consumers information at their fingertips before. This is going to change the immediacy of marketing while people are out and about. They are also going to demand deeper information from marketers at point of purchase or whenever they are thinking about a product line or entertainment vehicle. For instance, for opening weekend movies, mobile devices are vital for branding, getting plot info, etc. For phones that are equipped, you could provide the trailer, funny outtakes or cast/crew interviews and, more importantly, link to a movie ticket seller so they can buy the ticket from their phone.

Q. Why do you think Mobile Marketing will be an effective tool for connecting with moms?

A. Moms have zero time to waste. They are looking for solutions, comfort and information. If a marketer can provide this in a simple, effective and attractive mobile site, they will win with moms. Remember, no ugly technology has ever worked with a female audience—make it beautiful!

You need to catch consumers at the moment they become interested in your brand and nine times out of ten, especially if she's a mom, she's on the go.

Q. What does a company need to have in place to take advantage of the Mobile programs Hearst has in place?'

A. Excellent question, thanks for asking. They can create a campaign with us that includes buying banner ads across our mobile sites aimed at moms. If they have a mobile site already, these banners can link to that. If not, we can help them build that as part of our partnership within the Nokia Ad Network. They can email me at mobile@hearst.com for more information.

Twitter

"What are you doing now?" Are you sending "tweets"? If you aren't "twittering" from your mobile device, then you may be missing out on an emerging new marketing tool. It's called Twitter and users answer the very question I started this discussion with, "What are you doing now?" in 140 characters or less. The answer you send to your followers is your tweet. It sounds silly, time consuming and even far fetched but it is a tool that is catching on like wild fire and a great tool for connecting with moms.

Twitter is a social networking and micro-blogging site, which was founded in July 2006. It allows users to post updates that are distributed instantaneously to all their followers. The 140 or less character update can be posted via the web, text message or instant messaging. I am including Twitter in this chapter because many moms are using this tool via their mobile device. Users have their own profile page that displays their latest updates. In addition, users can become "friends" with one another, or simply be a "follower." Other than reading another person's profile page, a user can also receive others' updates through text messages, RSS feeds or third party applications. With Twitter you can make new friends or block people you don't want to talk to. You can also navigate through the site to find like-minded people. Twitter is a free service, although users may incur related text charges via their wireless company. I recommend visiting www.twitter.com and viewing the short introduction video. It's very informative and explains the entire concept of Twitter in less than 3 minutes.

There are other services similar to Twitter such as Jaiku, Pownce and Facebook. They have extended the micro-blogging capabilities of Twitter with functionalities that include blog posts, calendar sharing and music syndication. We will focus on Twitter for purposes of marketing with moms since so many mom bloggers currently occupy the Twittersphere.

Having explained the phenomenon of Twitter to many marketers, I know that the two immediate questions that come to mind are: "Who has time for all this blogging and micro-blogging?" and "Who cares about what a person is doing right now?" Well, I can assure you that the answers are the same for both these questions—Mom 3.0. She is not only spending time twittering, but she cares enough about others to follow what they are doing too. And to answer your next question, no, she's not a stay-at-home mom with too much time on her hands. She's actually a very active mom who is likely an influencer among her peers. Most likely a blogger who is keeping the dialogue with her audience fresh throughout the day in between her normal blog posts. This is yet another lesson marketers can learn from the consumer. Moms know that consistent communication is important to retaining her customers, which in this case are her readers, so she maintains an on-going dialogue with them by sending tweets.

Here's what a few "Twitterers" say about this new tool.

> "I use Twitter because it's a great way to keep up with what's going on in the blogosphere. Plus, if I ever have questions, all I have to do is ask and there are five to ten great responses in a matter of minutes." from **Jennifer James, Founder of the Mom Salon**

> "I joined Twitter in the first place because of peer pressure—I admit it! Every Mom in blogland seemed to be Twittering and they wanted to know why I didn't. So I joined. It's a great way to all "meet up" in one place any given time of day or night. You can Twitter from your desk or on the go from your phone. We tweet about what we're having for dinner, the weird lady in front of us at the grocery store, and we tweet our latest blog posts. We tweet about the latest news and it very well could be one of the best ways to keep people up-to-date with immediate news from all over the world! For example, the on-the-edge-of-their-seats waiting to find out my latest ultrasound results. Instantly

> everyone knew I was having a girl—just like they were there with me!" **from Stephanie Precourt, Mom Blogger, Adventures in Babywearing**

The earliest adopters of Twitter have largely been mom bloggers. Not only are they keeping readers interested in their life, but they are also using the tool as way to update family on their children and every-day activities. Connecting with family, friends or even coworkers is simple, easy and fun with Twitter.

Let's say Junior takes his first steps today. Instead of sending out a dozen separate emails to family announcing the milestone, the mom can now send out a 140-character message to everyone who is following her on Twitter. It's a solution that makes sharing easy for busy moms. The end result is that Mom 3.0 is actually saving time by leveraging the capabilities of Twitter.

It also puts mom in control of her time to socialize. Instead of getting into a long-winded conversation with a gabby friend, she can send out a tweet that lets multiple friends become aware what's on her mind. By reading the tweets of the friends she follows, she can catch up on what's going on with her friends when she has time. For momprenuers who are working from home, Twitter becomes a virtual water cooler that helps her feel less isolated during the day. Additionally, she can send out a tweet question to gather a needed resource or validate an emotion by sharing it with her group of friends, in 140 characters or less of course. It is reality television at its best to be able to know that another mom is watching the same program you are currently watching or that your friend in California is snacking on Ben & Jerry's too. It takes all the guilt away if you know that someone else is indulging with you.

Moms have been using Twitter in the way that marketers should be using it. They are leveraging the tool to market themselves, their blogs and products to other mothers. First, they grow their list of followers. They do this by posting buttons in places online where they already have a following or group of admirers. The button says something like, "Follow me on Twitter" and links the reader to her Twitter profile. Once there, the registered viewer selects to follow the mom. Each time someone new has selected to follow you, the user receives a notification email. The list of followers basically becomes your database who receives any marketing message you send them. Many mom bloggers, podcasters and vloggers will send out a tweet each time they post something new to their blog or upload a new podcast or vcast online. This allows her audience to conveniently click over to her blog, podcast or vcast. It's a great marketing tool to reach a large number of people all at once with updates about your product.

Twitter is a new media that moms are using to create solutions, simplify life and share information. As a follower of many twittering moms, I've witnessed how moms are using tweets to notify other mothers about a great sale, a new product they've discovered, debate political issues, root for their favorite hockey team and warn friends that they are going to be late to dinner. Some moms will issue questions to their followers in hopes of finding an answer to a parenting or business challenge.

I recently used Twitter to find a hotel in New York City with the help of my followers. As a way to nurture relationships, moms also share interesting articles with friends and family by including the link in their tweet. In order to condense the URL of the link so that they stay within the 140 character rule, many use TinyURL, www.tinyurl.com. It is a free service to make posting long URLs easier, and may only be used for actual URLs.

The premise that twittering is taking away time from other activities would be a logical deduction by outsiders who don't understand the Mom Market. The reality, however, is that moms are using Twitter, along with other wireless applications, to save time and communicate more efficiently with friends, family, coworkers and their market of consumers. It's time for marketers to take the lead from their consumers and mimic their behaviors to create a truly relevant dialogue.

Companies and Twitter

There are some companies who have seen the benefit of Twitter and are leveraging it to connect with their consumers. JetBlue is one of them. Here's what their tweets look like to their followers:

> @JETBLUE:
>
> @laughingsquid Thanks! Just wait until our new terminal at JFK opens! http://t508.com/ (Hope you had a great time in NYC)
>
> @BookingBuddy We're happy to support Runner's World, but don't worry, you can still channel surf and eat blue chips if you want!
>
> Two days left to win two trips on JetBlue to "The Simpsons Movie" premiere in LA. Enter now! http://tinyurl.com/2?6ql3
>
> The term 'Interwebular Chronicle' makes me laugh.
>
> View this month's selection of first-run movies from Fox InFlight Premium

JetBlue even used Twitter to explain why they engage in micro-blogging with their customers. From a post on May 9, 2008 at 1:23 P.M., they write: "Twitter matters because our customers matter. Brevity enforces honesty, and honesty breeds loyalty. The market IS a conversation."

There is a lot of truth in that brief statement. Sometimes we add so much marketing jargon around what we really want to communicate to our customers that by the time it reaches our market, the entire purpose of the dialogue is lost. Can you imagine if all marketing materials or messages were limited to 140 characters? There would be a lot more space on the page for the customers to react to, that's for sure.

Another big name on Twitter today is Dell Computers. The computer brand uses employees as their ambassadors to maintain a dialogue with customers. In fact, they have the largest number of employees in the Twitter network. They actually have several customer service people on Twitter who find complaints and address them. This proactive approach with customers goes a long way to creating loyalty, just as JetBlue attributed to in their above tweet. You can see why legal departments might be uncomfortable with this approach, however. It requires your company to forget about trying to use social media on your terms and use it as the customer uses it. Having an effective presence on Twitter means you have to interact directly with your customers in a public forum. However, if you want to relate with Mom 3.0 in a way that is natural to her, you'll have to conform to her means of communication.

Cautious marketers need only to remember a few key points to engage in leveraging Twitter or other online network tools. First, you must remember why moms use Twitter. They do so to meet like-minded moms, socialize and engage others in conversation, build their networks, drive traffic to their site, blog or product and share family updates. There's not a lot of room in these activities for marketers unless you are delivering useful, relevant and interesting information through your tweets. Your content cannot always be market driven. You have to treat Twitter the way Mom 3.0 is treating it. Communicate, share and socialize in a free and easy fashion. Reflect your audience by expressing humor, reality and sincerity. Most importantly, be honest and completely transparent to your audience.

The second recommendation is to put a face in front of the company. Moms are more likely to relate to another mother, and you most likely have a few walking the hallways or inhabiting offices. Engage a mom employee to be your face and maintain the dialogue the company. You may even find that there are several volunteers who not only enjoy socializing with outside moms, but also carry a real passion for sharing your company story.

Moving Forward with the MobiMom℠

The opportunities in mobile communication are numerous and will continue to grow as functionality increase in wireless devices. Now is the time to ready your company for the ability to market with Mom 3.0 on the move and as a MobiMom. Take the first step by entering the world of Twitter or other text based social networks. Become part of her mobile conversations by providing updates on service, product and topics relevant to her everyday life. Prepare yourself for the day that mobile applications allow you to speak with a mom when she's meal planning, juggling schedules or socializing with friends by creating natural points of entry in these behaviors. You may consider building wireless platforms on your own or partnering with others such as publishers, wireless providers or media companies. Apply a MobiMom mentality to anything you build between today and the day mobile reaches its potential. In other words, register .mobi URLs so you own them when you need them, build websites that are adaptable to wireless applications and create social networking platforms with text sharing capabilities.

There have been many technologies that have emerged and grown in recent history but none will be as welcomed to the Mom Market as the wireless explosion that is yet to come. Mom 3.0 will adopt and utilize mobile applications faster and sooner than any other consumer group because of its natural integration into their mobile lifestyle. Marketers who move forward with moms at the speed of wireless will reap the benefits of a deep relationship with their consumer. It means sales, retention and connecting with the lucrative Mom Market.

There is a whole world of moms out there just waiting to purchase your product or service.

Worth Mentioning

CHAPTER

THERE ARE SEVERAL TOPICS WORTH MENTIONING IN THE WORLD OF marketing with moms. If I were to write a chapter on each and discuss them with as much detail as I'd like, you would be holding a book the size of *War and Peace* in your hand. However since I feel strongly about empowering marketers with as many valuable insights as possible, I've included this catch-all chapter. I hope by briefly presenting these important aspects of marketing with moms that it will spark your interest to learn more at another point in time. Should you want additional information on any topic covered in this chapter, please feel to email me at Maria@bsmmedia.com. There are lots more insights where these came from and I'm happy to share them with you.

Ecosystem Selling™

Necessity *is* the mother of invention. If you look around at Mom 3.0 you see it everywhere. Moms inventing solutions for her everyday challenges whether they are processes that simplify her life or actual product inventions to modify the way she accomplishes chores. Some moms make money on their inventions such as Julie Aigner Clark of Baby Einstein or Allyson Shandley of The Tippy Cup. However, most of the systems and products that moms create go unnoticed to anyone other than her own family.

Every day moms create ecosystems of solutions. These are systems of product, technologies, processes and companies that she bridges together to help her get the job done. Any married male reading this book can relate to this explanation of female ecosystems. Picture this. You and your family are going on a road trip. You are in the driveway packing the car. Your wife comes out and immediately starts moving around things you've just placed in the trunk. You are thinking, "What am I doing wrong?" You aren't doing it wrong. You just aren't doing it her way. She's most likely already created an ecosystem of solutions for the long car ride. She wants the paper towels near the front so she can get to them when Jr. spills his juice box and the blanket needs to be in the second seat so that she can reach it when the dual air conditioning controls fail to evenly distribute the cool air to the warm side of the back seat. This is one instance where as her spouse you might not want to get into her ecosystem space, but as a marketer there are other opportunities.

Let's look at a mom creating an ecosystem as a consumer. The easiest illustration of this can be found in digital photography. We are going to speak more about the importance of digital photography in our next section but for now, recognize that picture taking is a huge touch point for moms as the Chief Family Archiver. A mom does a great deal with her photos. She stores them in albums to archive her family's history, uses them when required for her child's homework assignments, turns them into gifts during holiday periods and shares them with relatives and friend online. Ecosystems come into play when you ask mom how she prints the digital images she uses for all these tasks. I challenge you to find any 200 moms who give you the same answer. They will explain that for every intended purpose they have a different way to utilize the images on their photo card. The answer will often sound something like this, "Well for vacation pictures, I print them at Costco or Wal-Mart but for a homework assignment I use my home computer and if I want to order a photo mug for Father's Day, I upload a picture to Snapfish.com. If I want to store and share photos from a special occasion, I upload them to FlickR and if I want to include them in my blog I upload from my home computer." In this case, moms are creating an ecosystem of solutions for using their digital images but it goes on in the kitchen as well.

Ask a mom about her ecosystem for feeding her family and you will find an infinite number of solutions. In my house, it starts with a family dinner on Sunday night, a crock pot meal on Monday, a frozen casserole on Tuesday, leftovers on Wed and take-out on Thursday. Moms bridge together products, services and processes to create customized ecosystems of solutions. As a company, you need to be a part of the ecosystems she is creating. She's either going to do it with or without you so you might as well cash in on the established behavior of moms. When a company decides to provide moms the pieces that create the circle of solutions, I call that Ecosystem Selling. It's a concerted effort by a company to present the pieces to the puzzle that allows moms to create her solution system.

If we look back at the digital photography example, you will see that HP decided to become part of mom's ecosystem of solutions. Whether she needs to print her photos in a retail environment or online or at home for that late night homework assignment, HP has a solution to offer her. She can upload her photos to Snapfish.com to share them with friends or create gifts for family. She can use her compact printer in the kitchen to print a single photo while making dinner or she can insert her memory card into a kiosk while shopping at Walgreens or CVS and print her vacation shots. They have the tools in place for her to stay within their brand regardless of her photo ecosystem demands.

It seems logical that companies should sell to moms through ecosystems and indeed many companies have created bundled products as a tactic to do so. Most companies fall short in Ecosystem Selling because their marketing efforts operate in product silos. Marketing efforts rarely transcend departments, products or distribution channels. This type of execution is in direct contradiction to, not only the way moms do business, but in the way they maintain relationships. They don't see Nestle and see frozen dinners and formula in two different lenses. They see Nestle, a brand they trust to feed their baby and later a brand that feeds their teenager Hot Pockets. When they see HP, they see a brand they trust for printing documents at home so when it comes to printing photos they believe HP will deliver the same reliability.

It's only our marketing departments that divide up the relationship with mothers. As marketers we need to take our cue from our consumers. They want easy and simple, not only in your products, but in the way they do business with you. They want a simple way to carry on a relationship with your brand. Companies need to think like mothers and examine how they are using your products or services. If you can't provide every solution in the ecosystem then you should at least be one piece of it.

I'll give you one more example. A few years ago I was invited to a meeting with a container company. They wanted to discuss marketing their products to mothers. In preparation I did some research to find out how moms were already using their products. At the meeting, I spoke about moms using containers to sort scrapbooking supplies, to organize Legos, and to send crayons to school. Referring to Ecosystem Selling, I pitched all the opportunities for marketing to the roles containers played in mom's daily life outside of meal planning. The strategy was to sell to her needs and be the container in her ecosystems of solutions. Much to my dismay, the male product team refused to believe that moms were using their products for anything other than leftovers. What a shame to leave so much opportunity on the table.

Social Selling

No other business strategy puts moms more in control than social selling, otherwise known as direct sales. Such an approach to the Mom Market has built billion dollar companies such as Avon and Tupperware. Today the direct sales industry enjoys tremendous growth. According to the Direct Selling Association (DSA) retail sales of more than $28 billion for the direct-selling industry occurred in 2002—and employed an estimated 13 million independent direct salespeople across the country. This is an increase from about $22 billion in 1997, with 9.3 million salespeople. A large part of the growth in this industry is being fueled by entrepreneurial moms and companies that never before considered direct sales as part of their business model.

Today non-traditional direct sales opportunities from baskets to bathing suits are springing up. Even Jockey, the long time underwear brand, offers moms the chance to host living room Jockey parties where they may view and purchase product. It may be a strategy that your company might explore as a way not only to drive sales, but connect with moms.

So why are moms drawn to direct sales? Well, they aren't. They are attracted to social selling; a description I feel fits the motivations of moms far better than the industry buzz word. I've studied moms and business since the late 1990's. I went on to publish a book on the subject titled *The Women's Home-based Business Book of Answers* (Prima, 2001). This was the first time I asked mom business owners about their motivation in running their own company. The answer may surprise you. Most moms are not, and I repeat, not, motivated by money. In fact, I've asked this question repeatedly year after year and each time gather the same result. Approximately 85% of moms who own businesses say that money is not their primary motivation.

Among the top reasons for getting into business are "to get out of the house," "to have an interest of my own" and "I saw an opportunity to share an idea, service or product" with other moms. Isn't it amazing how closely aligned these are with the five core values of moms and their key motivators? They are attracted to direct sales because they can nurture relationships with friends outside their home, share valuable information with other mothers and demonstrate accomplishment.

A few years ago, I was working with Highlights Jigsaw, the direct sales arm of *Highlights* Magazine. Their direct sales business focused on educational toys that were unique to the United States. Moms were doing quite well hosting living room toy parties with their peers and selling product for Highlights along the way. I will never forget asking a group of mom distributors why they had selected Jigsaw's toys over candles, soaps or baskets. The explanation I heard repeatedly was that they enjoyed teaching moms how to use toys in a way that enriched their child's play. They actually saw their job, not as toy saleswoman, but as an educator. Several moms even used the word "enrichment" when describing their Jigsaw business as a way to enhance the lives of fellow moms and their children.

When I drive the discussion toward money and business, moms will tell me that their direct sales business enables them to pay for the extra luxuries at home such as after school activities and vacations. I can't tell you how often a mom will tell me that she sells vitamins, toys or candles in order to socialize with friends while earning a little extra income. Direct sales work with moms because it allows her to multi-task.

Creating a social selling program is a viable business model for almost every company that currently sells product online. You most likely have a distribution system in place that can be modified to process orders submitted by your offline sales team of moms. You also probably have marketing materials that can be easily turned into catalog pages or brochures. All you are missing is a group of moms who will embrace your product so much that they want to share the news with all their friends and family. I realize I am simplifying the process but I do so in order to help you see it's not impossible from an investment standpoint.

It does take management time and resources, creation of financial models to pay out commissions and rewards; however it's all doable. For smaller companies, it's a great way to expand your workforce very quickly. Hopefully I've convinced you of the potential but I'm sure you are asking the obvious question—"Where do I find the moms?" I get this question a lot whenever I'm conducting sales training sessions within direct sales companies. It's the hardest part of direct sales just as it is in any line of business—recruiting good people. The answer lies in the life stage of the mom.

You have to target moms who are at a changing point in their life stage. My best advice to direct sales recruiters is to comb the population of kindergarten and first grade moms in your community. These moms suddenly have six to eight newly-found hours in their day and it's normally at this change point that they seek a way to fill those hours. Another change point is when her child is 6 months old. If she has returned to her old job, most likely on the baby's 3 month birthday, she is likely either debating her choice or seeking an income generator that allows her more flexibility in her schedule. Of course, this does not occur in all moms but it seems to be the timing for the great work debate. I attribute it to the fact that at 6 months her baby is reacting more to her presence and it is at this point that she is comfortable with her new role as a mom. Another change point for moms occurs when their children leave the nest for college or a career. In fact, empty-nester moms present an ideal market for companies looking to expand their sales force with mothers.

Although social selling presents a wonderful way to connect with moms and drive sales to the bottom line, it does come with its share of challenges. The first is the double-edged sword of their motivation toward money. Some might say it's great that moms aren't motivated by the almighty dollar, however for most corporate cultures, this is a contradiction of terms. Management does a good job, they receive a bonus. Share prices go up, shareholders receive a dividend. Money is the commonly used form of corporate motivation. It takes an innovative sales manager to keep moms motivated without throwing dollars at her. You have to recognize what drives her to devote her time to your brand and respond with the appropriate reward. Once you make this point of connection however, she will reward you with incremental sales to your bottom line.

Digital Photography

You may think that you can skip the next few pages as I discuss digital photography because you aren't in the camera, printing or gifting business, however I would caution you not to skip this topic. Photography is perhaps the most influential, motivational and frequent elective behavior a mother demonstrates. I call it elective because unlike feeding a baby or clothing children, it has no negative physical effect on a human being if it's not performed. As the Chief Memory Officer of her family, moms embrace photography. Consider some of the statistics that BSM Media has uncovered about moms and photography. Over 80% of photos taken are taken by moms and 58% of moms have published photos privately or publicly. Mothers begin to think of their holiday photo cards as early as June 1. Eight out of 10 moms have done some type of scrapbooking in the last year driving the scrapbooking industry to over $2 billion in sales.

Digital photography has fueled the influence of archiving memories by expanding mom's ability to connect, share and nurture. I would argue that the average mother spends more time involved in some type of photo activity over her adult lifetime than she does engaged in exercise. Moms love photos because they allow her to archive her family's journey as well as share experiences with friends and relatives. Ironically, they also give her a sense of accomplishment and position her family as a showpiece. Photos allow her to brag about her children without really bragging.

Here's an example of this that I attribute to the friendly competition that moms exhibit among their peer groups. Think about the holiday season. Photo cards are very popular. You receive them with cute bows tied to the top or matching envelopes but it's the photo that I want to focus on. Each year, I get at least one card with a beautiful picture of a family on it. The children are all dressed in matching outfits, hair neatly tied back in big bows and everyone is smiling at the camera. As if that's not enough, the card arrives at your home the day after Thanksgiving. This is an example of using photography to not only showcase the family but to demonstrate her personal skills and sense of accomplishment. The message the sender is delivering to you is, "Look, I was able to buy matching outfits, get my kids dressed, smiling and all looking at the camera at once and I did it before the end of November." Do you know the power this yields in the world of moms? The sender's position is likely to be elevated among her hierarchy of peers and for the next eleven months she will enjoy the benefit of her labors.

The introduction of digital photography has been a blessing and a curse for moms. It's easier than ever for her to capture and print images of her family whether she is using a digital camera, Flip camera or cell phone. However with the ability to capture more photos comes the challenge of printing, organizing, sharing and creating keepsakes with more photos. In fact, Hewlett-Packard estimates that about 86% of all digital pictures never even leave the camera. What digital has done to photography is instill a sense of "disposable" to picture taking.

In the old days of film, a mom would take a picture of her children while on vacation and hope it turns out the way she intended it. Today, she may take two or three shots of the same situation, reviewing the image before moving on. She will go through a second selection process when she selects which images to upload or print. Gone are the days when she dropped off the film at the drugstore, leafed through the finished photos and sold back the ones she didn't like. Today she is in almost autonomous control of her archiving abilities. In fact, camera retailers will tell you that twenty years ago, the family camera or video camera purchase was largely made by the man of the house. Think about your own father's role in this selection when you were growing up. If you are Boomer, your dad probably had a big fat camera bag on his shoulders whenever you traveled as a family. Today the camera purchase is largely controlled by the mom of the house. If she's a blogger she most likely not only has a digital camera but a Flip video camera as well. A Flip is an easy-to-use pocket video camera that plugs directly into your USB port and uploads video to your computer for use on blogs, vlogs and social network sites.

The part of photography that hasn't evolved is the transferring of digital images into hard-copy prints, and sharing or archiving them. When they do find their way into print, the photos end up in boxes under a bed waiting for the day when they can be placed into an album or mailed off to grandma. In fact, according to my research, photo management is one of mom's top frustrations.

While digital photography has become a challenge for mothers because of the increased images she is capturing, technology today is allowing her to do more with those memories. Today she can share them with friends and family via gaming systems, upload them to Snapfish.com and turn them into gifts like calendars and coffee cups, insert them into electronic scrapbook templates and even put them into hard cover bound photo books with or without words. With all these capabilities, she can more easily share photos, nurture relationships through those photos and engage in friendly competition almost 24/7—whenever the time is right for her.

The ability to do more with her photos is so strong that I see a bigger marketing opportunity for the Nintendo Wii beyond family gaming. I believe that one of the most marketable features of the game system for mothers is the ability it has to play photo slide shows and cross-country viewing of the same show. It's so simple to do with the Wii. A mom merely plugs in her memory card and her television is suddenly turned into a giant slide show screen. Haven't we come a long way since the bed sheet on the wall our fathers used to show family videos?

A mom could have pictures of her birthday girl playing while entertaining family at her daughter's birthday party. At an anniversary party, the hostess could have photos flashing continuously on her living room television, creating a unique source of entertainment for the guests. It's an idea that I'm sure they've thought of; however, as in most technology companies, I am sure the marketing messages are governed by engineers proud of the primary functionality they have created. It is not an uncommon mistake made by technology, whose pride for their product normally resides in the newest features rather than the benefit to the consumer. As in the case of the Wii, consumers hear a great deal about the unique wireless nunchucks that allow the player to move along with the game avatars rather than the benefit that it's multifunctional for the entire family. The photo sharing ability of the Wii would be a mom benefit I would definitely include in all marketing materials for the game system.

Leveraging the Power of Digital Photography

As the Chief Family Archiver and Historian, moms are looking for solutions that are easy and fun. This creates a great marketing opportunity for companies and the good news is that you don't have to be Kodak to take advantage of them. It's my belief that just about any company can leverage the power of photography to connect with moms. For instance, if you are a food company your products, and all the things a mom can do with your products, can make great photo marketing opportunities. Allowing mothers to submit photos showing how she is using your product for your website can go a long way to establishing a connection with her. It also allows you to incorporate peer-to-peer marketing into your online strategy. The other benefit to you as a marketer is that her natural tendency to share and brag kicks in and soon she is sending all her friends to your site to see her pictures showcased alongside your products. The viral effect starts to take place.

Digital photography is also a big part of the blogging world. Mom bloggers love to include photos of food, places and product within their posts. Make certain that you include digital images of your product whenever contacting a blogger. There's nothing like a mom influencer or a blogger flashing your product pictures to her audience. As a non-photo based business, you can identify opportunities by examining how moms are using photos and fold yourself into her existing behavior. It's very simple. She's using photos to share, archive memories, brag, demonstrate accomplishment and skill, nurture relationships, gifting and showcase her family. Now, how you can get involved in one of those activities? One way might be to expand your product design to incorporate personal photos, as Kleenex has done.

In 2007, the tissue maker found a way to generate a new revenue stream and incremental sales by leveraging the power of photos. Kleenex now offers customized photo boxes at www.mykleenex.com. For less than $5.00 per box, Kleenex invites consumer to get creative and design their own oval box of tissues with their favorite photos. They are great ideas for weddings, birthday parties and other special occasions. It's a great way for Kleenex to connect with the moms through their desire to customize product, show off their children and engage in friendly competition. A customized box of Kleenex certainly says this mom has found a new way to elevate herself above others. Without being a photo-related business, Kleenex has successfully found a way to create a new revenue stream by leveraging mom's love of photos.

There are thousands of ideas of how a marketer can leverage a mom's love of photos to establish a dialogue with her and create viral marketing campaigns. Entertain my passion for the subject for just a few minutes as I share more ideas with you. First, let's go back to the holiday photo card I discussed earlier in this chapter. Knowing how important it is for a mom to capture the perfect holiday shot of her child, consider how you might be her solution. If you are a clothing retailer, you could make the offer to moms to visit your store on a particular day in November and dress her child up in any outfit in the store. You'll then have a photographer on site to capture the child's dressed-up look. I assure you can find a great freelance photographer simply by placing an ad on Craigslist. You print the photo with a desktop printer, the mom leaves with a picture and, I will bet, the outfit her child is wearing in it. Your promotion entices her into the store because it presents a solution to one of her greatest seasonal challenges and allows her to multi-task. She gets her holiday photo while she shops. Just about any retailer can do a promotion similar to this. If you are a grocery store, you can create a back drop with flowers for the photo. Last December, Wal-Mart took photos with Santa in the garden department nationwide using the Big Guy and seasonal flowers.

Other ideas include photo contests that feature your product or moms engaging with your brand. Baby photo contests are always popular and allow you to generate a rich database of those moms who enter. Every mom thinks they have the most beautiful child in the world and have little intimidation of sharing a photo with you to prove it. Whatever type of photo promotion you decide to launch, remember to make it easy and incorporate a platform or tool for mom to share the experience with others.

Gaming

Another topic worth mentioning is gaming. Moms are engaged in gaming, not only as the purchaser of games for her child, but as a player herself. In fact, according to the Entertainment Software Association, 38% of video gamers are women playing up to seven hours a week. The average age of female gamers is 33, while the average age of game buyers is 40, confirming that many of these women are parents purchasing games for their children.[1] Game designers are taking notice of this and creating games and gaming systems that will appeal to moms and females in general. A trip down the aisle at Target or Best Buy will introduce you to titles such as Hannah Montana Music Jam which can be played on a pink Nintendo DS. But it's not only the titles and color that sell the product to moms, it's how it physically and emotionally fits into their lifestyle. Many have credited the size of the Nintendo DS and the fact that it can comfortably fit into the smaller hand of a woman with its success in the female market. Furthermore, games like Sudoku and MindTeaser are also used by older women to relieve stress while strengthening their minds. It's multi-tasking at its best for the best multi-taskers.

Perhaps the company who has most successfully tapped the appeal of gaming and the video game purchasing power of moms is Nintendo with the launch of the Wii. The company's smartest move was to conduct boot camps to teach moms about video games and allow them to try it out for themselves. They also enlisted the help of mom influencers who hosted Wii parties for the company and invited their friends to attend. In the end, Nintendo was able to position gaming as a family activity and moms loved it. The benefit for mom was that the Wii brought her otherwise isolated tween or teen into the living room providing a means of communication between the two of them. My research shows that most moms who

1 Jean, Sheryl. Retailers and Manufacturers Starting to Target Female Players, *Star-Telegram*, www.star-telegram.com (accessed November 21st, 2007)

are casual gamers do so to relax. What's interesting, however, is that even in relaxing, they are multi-tasking. Gaming moms select titles which give them mental exercise, such as Brain Age and other word or puzzle games. While they are eager to gain mental exercise, many of these women are gaming more than they are physically exercising according to an AOL survey. The creation of women-friendly titles and gaming systems has paid off for Nintendo. Females now account for 22% of Nintendo DS handheld games, which is double the number of women who bought GameBoys just a few years ago.[2]

Moms have an interesting opinion about gaming and it changes based on their level of engagement—either solo or with their family. When we ask a mother about her own game playing, she will tell you that it allows her to escape for a few minutes during the day or stimulates her mind when she needs a break from her everyday chores. However if you ask her about your child's gaming, she will often respond verbally while making the thumbing motion that imitates her child on a controller. Although she is accepting of her child's gaming, she will express concern for the isolation and alienation it creates between her child and the family. We see more and more moms making the effort to bridge the gap between them and their children by learning how to play their kids' games.

Virtual Worlds

Another form of gaming is engagement in virtual worlds. These online communities are becoming more and more popular with not only moms but her children. The growth of sites such as Webkinz, Club Penguin and Second Life has made it easy and fun for a mom and her children to play within virtual communities. Moms become target consumers; not only for the virtual worlds where she plays, but also for the sites her child visits because of the role she plays as gatekeeper for her children. Let's examine the three types of virtual world engagement that exist with today's Mom 3.0.

2 Ibid

Virtual worlds allow moms to escape the realities of her day-to-day life. Although almost 8 million adults take part in sites like Second Life, I estimate that the mom number is probably somewhere around three million. The mom who possesses an avatar and goes online to Second Life is most likely in her mid-thirties and she does it late at night. In fact, our research points most specifically from 10:00 p.m. to the midnight hour. Certainly there are opportunities for brands to interact with her during this time through product placement, online sponsorships and corporate sponsored avatars.

The more frequent occurrence of virtual world interaction can be found in the virtual worlds her child visits. Most of these sites have varied levels of engagement, starting with free play but opening up to expanded play as you pay. Mom's interaction normally begins when her child asks her for a credit card to enroll. Sites such as Club Penguin, which is now owned by Disney, and Barbie Girls charge anywhere from $4-$9 a month for children. This pay-for-play or subscriber type gaming is a growing trend among virtual gaming worlds. However, even with the growth of virtual worlds, more money is still being spent on physical gaming.

According to the NPD group, kids are still more interested in physical media such as DVDs and videogames; spending $13-18 per month while only spending $6-$12 a month on digital content. In fact, it is estimated that Club Penguin has 4.5 million unique players according to comScore. The other model for virtual worlds is product based. Although you are not paying a subscription rate to play the best games on the site, the child (or mom) must purchase product that gives them access to various levels of play. BSM Media research shows that 85% of mothers claim they know where their child is going online. However, even this knowledge does not make them a fan of video games or online play. Nearly two thirds of parents worry about their children participating in online communities.[3] According to the same survey, 25% of parents believe

3 Youth Markets Alert Publication, "Top Parental Concerns about the Internet," *Center for Digital Future, USC Annenberg School for Communication* (February 15th, 2008.)

their children spend too much time online, a number that has increased since 2000. The greatest concern is online predators; second only to the concern that the Internet limits real-world interactions with friends and family.

Marketers can leverage their relationship with moms by speaking to her concerns as they attempt to win over her child. Referring again to research conducted by my team at BSM Media, moms say they want to limit their child's ability to chat with other players, understand the rules of engagement and know that the company behind the virtual world shares their values. I always recommend a mom's page on any virtual world or website for children under 12 years of age. Some sites have tested timers throughout the years, but I've never known moms to specifically select a site because it limits the child's time online. More often, I see a mother's tolerance level increase as her level of confidence with the brand increases.

There's an irony in a mother's fear of stranger danger in virtual worlds. She says she worries about the online community her child plays in, yet she doesn't invest the same due diligence she does in the physical world. If her daughter, Kate, asks her to take her to a movie with five friends, she is likely to ask who else is going, what time Kate will be home and the location of the theatre. When it comes to virtual worlds, however, few mothers apply the same level of scrutiny that they apply in the physical world. They will ask other mothers about their familiarity with the game or base their opinion on their level of trust with the brand who owns the platform. For companies with high brand affinity by moms, a virtual world for her child may be a winning strategy toward expanding your current market.

One company who has created a successful virtual world is Precious Moments. We spoke earlier in the book about their strategy to engage a younger population of customers. The tactic they deployed was launching a virtual game world for children. Their goal was to attract the daughters of the moms who already trusted their brand

and introduce them to a new, younger and hipper line of products. They built an entertaining and engaging online destination for children, and they took it one step further in leveraging their existing relationships with moms. They built elements into The Precious Girls Club, www.preciousgirlsclub.com, that allow a mother to interact with her child in the virtual world. For instance, a mom can send an email to her daughter's computer inside her virtual bedroom. To keep mom informed on her daughter's level of play within The Precious Girls Club, emails are sent back to the mother with play updates. The final, and perhaps most unique, feature Precious Moments created along with the virtual world was offline activities that moms and daughters can do together. These offline activity sheets are found throughout the virtual world, downloaded and printed for play in the physical world. It plays directly into the fear of moms that video games alienate her child from family interaction. In the spirit of full disclosure, my company, BSM Media, created all of the mom and child interactive elements for the site.

For the family led by Mom 3.0, I believe the next generation of gaming and virtual worlds will require some level of family engagement. There will be games that allow each member to personalize and customize functionality including player interaction, characters and avatars and sharing capabilities. Additionally, mom and her children will be able to play the game in universal delivery systems that include playing the same game via hand-held device, minivan DVD, computer or TV. Although I do not believe that the pay-for-play model will survive, I am inclined to think that the product model will continue to grow. I think the latter is more acceptable to moms because their child obtains something tangible from the relationship with the virtual world. In other words, she doesn't mind paying $10.00 for a stuffed animal that enhances her child's play online because when the child tires of the website, they still have a toy to show for the relationship.

Marketing to Moms through Her Children

We couldn't examine marketing to moms without including her children in the discussion. As the Chief Financial Manager of her family, mothers are the gatekeepers of, not only household purchases, but often the final word on her child's spending. Marketing through a mother's toddlers, tweens and/or teens is an effective way to capture the buying power of the Mom Market. It is a successful marketing strategy because it can be used with any stage of motherhood. It plays to a mother's desire to make her children happy. This is why marketers have long used the nag factor to gain mom's approval to purchase a product.

All mothers want to please their children. Today's moms aspire to give their children more than they had as children themselves. The definition of "more" takes the form of material things from family vacations to the latest computer to the newest style of Nike shoes. Moms who overindulge their children with material objects have been termed "entitlement moms" by author and parent educator Karen Deerwester. Entitlement moms feel that their children are in some way entitled to more than she had as a child. It's a popular mindset for many Boomer and Generation X mothers who waited to start a family and established their financial stability while doing so. These are the moms who are escorting their six-year-olds to pedicures and hosting spa parties for their tweens.

Even those mothers who do not fall into the entitlement parenting style have a difficult time saying "no" to $100 tennis shoes or $85 American Girl dolls. The pop culture marketplace has created an arena that is associated with a child's level of esteem and peer acceptance. A mom will give in to her child's request in fear that she will be the reason for her child's alienation from her/his peers because her child is the only one without those shoes. The fear of creating a playground outcast or teen loner because of the absence of the latest acceptable fashion logo is too great a burden for a mother to bear.

The final reason that marketing to moms through their children works emerges from a recent desire by mothers to find balance in their busy lives. Time and energy starved mothers today pick the battles they fight with their children in order to obtain some type of home/life balance. Fighting with a straight-A teen over the brand of their jeans can waste precious time and energy. Instead of fighting this battle, a mother may concede to letting her wear the type of jeans she wants as long as her grades remain high.

Make no mistake about it, the child market offers advertisers an incredible opportunity in itself. In size alone it offers a desirable market. According to the 2000 Census Bureau survey, there are 80 million Americans under the age of 18, and 39 million kids ages 5-14.[4] James McNeal, who has spent over 35 years studying kids and spending, estimates that children influence over $300 billion in their parent's spending annually. Households with school-aged children outspend households without children by at least one-third. [McNeal, 1999] "Tweens," kids between the ages of eight and fourteen, are the largest demographic group among children today. Marketers are attracted to the child market because it offers the opportunity to gain what we used to call at AutoNation, "a customer for life." Companies see children as a customer today and in the future. Many child marketers call it the "cradle to grave" market. They win the child's admiration as a youngster when he can influence his parents' buying decisions, later he becomes a customer when he is spending his own money and much later when he purchases the product for his own family.

The irony is that the same companies who see children as potential customers for life forget to incorporate retention programs later in the lifecycle of the consumer. Newspapers spend millions of dollars introducing children to reading newspapers through current Newspaper in Education programs. Their goal is to get youngsters in

4 U.S Bureau of Census, *Population projections of the United States by Age, Sex and Race: 1995 to 2050* (Washington, D.C.: GPO, 2000), http://www.census.gov.

the habit of reading a paper every day. In the 90's, many dailies introduced features such as teen pages, family sections and children stories yet they failed to implement retention programs for adults. Smart companies will realize that unlike Peter Pan, children grow up and they must refocus their marketing efforts.

In the world of Mom 3.0, children contribute even more influence over the purchases of moms. Through the use of text messaging and other wireless communication, mothers are gaining their children's input on everything from cereals to chips while they are in the aisle. Today's mom takes pride in using technology to bridge her family together in not only scheduling, but in validating her buying decisions. Marketers who are trying to sell products designed for children should empower their young consumer with the right messages to deliver to the gatekeeper.

One of the best marketing strategies I've seen in this area was created by MCI, the wireless company. They actually created a PowerPoint presentation with all the reasons a moms should give her child a cell phone. It spoke to safety and enhanced communication. They made it available online to tweens and teens with the idea that they could present the most effective case possible to their parents. It was a brilliant way to have children deliver your message verbatim.

Disney's ToonTown did something similar a few years ago. They gave children a free subscription to their site. A day before the subscription expired, the child was able to create a customized letter to their mom. The letter explained what a great time the child was having on Disney's ToonTown and, by the way, the good times end tomorrow without paying the subscription fee. The goal of marketers with a dual consumer should be to educate the mom on the benefits of the product while selling the features to her child.

A Global View

As the world of consumerism shrinks, more and more companies are looking at the world globally. What they are finding as they look beyond the US borders are rich mom markets. According to Euromonitor International's publication, "Who Buys What," which analyzes expenditures in over 70 markets and 35 countries, women influence 80% of *all* household spending. The growth of female consumers' power has largely been fueled by more women entering the work force around the globe. Your company may not be ready to sell product to Asian mothers or service the moms of Costa Rica, however I thought a brief look at mothers from a global prospective would be beneficial. The topic could actually fill an entire book so I'll just focus on the major differences and more importantly the similarities with international moms and US mothers.

The greatest variation among moms globally is the point of entry into motherhood. In the US the number hovers around 25.1 years old, according to the National Center for Health Statistics. Remember, however, that this number is expected to go down as Generation Y women have their first baby at a younger age. The overall birthrate in Europe has gone down as women's participation in the workforce has increased. Europe has seen the number increasing over the past decade. In England, the average age of first time births has gone from 22 to 24 since the 1970's.[5] France has also seen a rise in age. The average French mother will have her first baby at 26 years old, compared to 22 just ten years ago.[6] These numbers are being driven by the overall first-time marriage rate increasing in the European block. In Hispanic countries, the age of first time births is slightly lower than Europe. Women in these countries typically enter motherhood around 22 years old.[7]

5 European Council (1994), Recent Demographic Tendencies in Europe, 1993, European Council, Strasbourg.
6 Ibid
7 National Center for Health Statistics. "*Mean Age of Mother 1970-2000.*" www.cdc.gov. (accessed June 10th, 2008.)

The influence of religion and culture play a part in keeping this number low. Interestingly, China has one of the highest first time birth ages in the world, with the average age of women entering motherhood being 30 years old. This number, however, fluctuates greatly between women living in populated cities such as Shanghai and rural areas. Women in larger cities are taking advantage of the professional opportunities opening up to them and prolonging their entry into marriage and motherhood. Not surprisingly, the financial independence many of these women are discovering is also driving up the number of single mothers in China—a cultural shift that I am sure will create a series of paradigm shifts in the future. Another reason for the high first time birth rate in China has to do with a woman's government-regulated ability to have more than one child. Many Chinese women will wait for the most optimum time to have their one and only baby. There is no rush to give birth if you know that other children will not follow. In a sense the biological clock has been modified by the Chinese government.

The global expansion of the workforce has affected the Mom Market in ways well beyond the age of first time births. It is influencing sales of products and the way in which marketers must approach a growing market that is engaged in balancing work and family. For some companies, the growing number of mothers in the workforce has yielded unexpected growth. For instance, sales of Nintendo handheld game systems in Asia have been linked to women purchasing the product as a way to pass the time while commuting to work by train. In the same way, advancements in some mobile devices in Japan have been attributed to mothers using wireless functionalities to keep up with their children while they are working. In markets such as Canada and Norway, where the percentage of working mothers exceeds 60%, companies need to consider products that help them bring simplicity to their hectic lives.

Another consideration when examining global Mom Markets is the influence of a mom's own mother on her buying decisions. BSM

Media research shows that Anglo mothers typically seek the advice of her peers after the birth of her baby. We've attributed this mom-to-mom influence to her desire to customize motherhood and do things "different" than her mother did with her. We hear this a great deal from moms. "I love my mother but I want to do things different with my baby." In contrast, mothers with Asian or Hispanic heritage seek their own mother's advice more often when it comes to parenting and product suggestions. We see this a great deal when it comes to food brands and consumer packaged goods.

Although there are social, economic and cultural differences that exist within the global Mom Market, there are commonalities among moms that can be leveraged by marketers. My family will tell you that I am always studying the behaviors of mothers. No matter where I am, on a plane, in line at the grocery store or in public bathrooms, I am watching and learning. As I mentioned in Chapter Two, I've spent a lot of time observing families in dining halls at Disneyland Paris. Almost each and every mother would perform the same behavior in feeding her children. She would help the child sit down at the table, then she would examine the child's plate almost as if to confirm they had food to eat, and then, begin to eat only after she felt confident her child's needs had been met. Yes, the foods they selected from the breakfast buffet were quite varied, but their behaviors as mothers were universal. No matter where their place of origin, I visually confirmed that moms have an innate sense of nurturing their offspring and consistently put the needs of their child before their own. It's an illustration of how the five core values we discussed in Chapter One transcend the globe. Moms everywhere possess some level of engagement with these values. To refresh your memory they are: 1) Health and Safety; 2) Family Enrichment; 3) Value; 4) Simplification; and 5) Time Management. These are strong messages that can be leveraged around the globe and, by overlaying cultural elements to them, can help you connect with mothers no matter where they live. There is a whole world of moms out there just waiting to purchase your product or service.

Marketing to Moms Coalition

Chances are if you are reading this book, you are engaged in selling to moms, a task more easily done with constant market insights. I encourage you to join the Marketing to Moms Coalition, www.marketingtomomscoalition.org in order to stay on top of the latest research. The Marketing to Moms Coalition is the only industry group dedicated to furthering an understanding of America's most powerful consumers. The goal of the group is to share knowledge and insights about Moms to help marketers create programs that truly engage moms. The organization was started in 2005 by a group of Veteran marketers in the Mom Market who were determined to create a platform for sharing best practices. The founding members were Bridget Brennan of Speaking Female, Teri Lucie Thompson, who at the time was with State Farm and now works with Purdue University, Amy Colton of Current Lifestyle Marketing, Michel Clements of The Insight Group and myself. The Board of Advisory includes Mary Dillion, CMO of McDonald's, Becky Chao of Nestle, Robert S. Matteucii, CEO of Evenflo, Katherine Durham of Hewlett-Packard and Nora L. Linville of American Airlines.

Membership is free, although group sponsored events sometimes carry a small fee to attend. In addition to releasing research, the Marketing to Moms Coalition sponsors the HER award, an annual presentation to a company whose marketing campaign honors, empowers and respects moms as a consumer. Past recipients of the HER award include Whirlpool and Sara Lee. I highly recommend joining the Marketing to Moms Coalition as a follow up to this book.

The integration of the old and new will allow you to connect with Mom 3.0 on a broad level as well as intimately online.

Conclusion

CHAPTER

FROM DIAPERS TO DESKTOPS, MOTHERS ARE PURCHASING MORE product than ever and through blogging to vlogging, they are telling more moms about these purchases. In the previous chapters we discussed both the spending power of today's mom consumer and the new ways they are spreading the word through peer interactions, social networks and new media. My hope is that you will take this new found knowledge and integrate it into your marketing strategy to successfully connect with this powerful market. I have no doubt that the right connection will generate increased brand awareness, incremental sales and customer loyalty. The secret of reaching these goals can be found, of course, in the execution of programs created with your knowledge of Mom 3.0. How does a marketer living in a Web 2.0 world with shrinking budgets, demand for measurable results, desire to try new initiatives but a need to continue leveraging traditional media begin marketing with Mom 3.0? This is what I'd like to tackle in the final chapter of this book. I want to give you the ability to see the opportunities that exist for integrating Mom 3.0 into your marketing 2.0 budgets and strategy without causing your management and legal department to question your sanity. It's a big task for one chapter but an important part of tying it all together for you.

During the course of writing this book, a reporter called to ask me for a pre-release interview for *Mom 3.0.* I graciously accepted but was surprised to learn from his questions that I had omitted one very key point in the discussion of today's mom. It's one that may help bring clarity to the tactical application of marketing with moms. The reporter asked me to distinguish between Mom 2.0 and Mom 3.0. Great question! What's different about her? The question drew a moment of silence from me. The difference between Mom 2.0 and Mom 3.0 can be found in the power she possess in determining the "what" of what she does with your marketing message once she's received it.

In a 2.0 world, marketers incorporate an element of interactivity in their brand messages. They ask her to fill out a poll or they provide her with product coupons to share with friends. It works and is producing results for many companies. The difference in the 3.0 mother is that this mom receives your marketing messages and decides "what" she's going to do with it. She has a sense of empowerment that she can affect the outcome of your marketing with her own interpretation of it. You may send her a plastic container for leftovers along with a brochure that speaks to the new features you've added like colored lids or collapsible bowls. Mom 2.0 may tell other moms via a playgroup or message board about the pretty purple lid but Mom 3.0 will blog about the dangerous polymers that are generated if the bowl is put in the microwave. The latter takes your product, educates herself with the materials you provide but goes one step further in determining what "she" wants to say about it and how. The 3.0 Mom recognizes that she has the ability to impact change, whether that's to sell product or educate peers.

All of this makes the "with" in marketing with moms all that more important. Marketers need to develop a relationship with Mom 3.0 in order to engage her and your purse strings. All of this is actually very good news for the marketing professional we described earlier in this chapter. Budgets for social media and online media tours with bloggers are far less expensive than traditional print advertising, yet compliment these more mainstream media buys quite well. The integration of the old and new will allow you to connect with Mom 3.0 on a broad level as well as intimately online. It also makes the shift or change in your marketing strategy palatable for almost any corporate culture.

Let's look at tying together all the elements of marketing with Mom 3.0 in various corporate environments. First I'll address the most common scenario—a major global brand who knows instinctively they need to shift some of their marketing efforts toward Mom 3.0. They read about Mommy Bloggers in the news almost daily, they see the viral effects of YouTube and know it's the right thing to do. Brand managers and media buyers place television ads for the greatest reach and place print ads in the appropriate parenting publications. For online efforts, they need numbers so they make media buys on the largest mom websites or on the parenting channels of women sites like iVillage. If you ask them what new types of interactive marketing they are doing for their brand, they will often point to an online contest or an outreach sponsorship with a PTO organization. These are all really great marketing tools; however, they lack the intimacy of a relationship with moms. As we've learned already, the relationship is the strongest link in the chain effect to success. The marketing plan needs to be infused with some of the initiatives we've discussed previously in this book.

I find that a good place to start is with bloggers. Everyone loves to see results and the most visible way to show your manager increased visibility in the marketplace is with mommy bloggers. Select a small number, like ten, or work with a company who has established relationships with these women. Clearly define your goal, whether its brand awareness or creating buzz, and make sure your efforts support it. I suggest that for first timers, the easiest benchmarks to quantify are exposure and reach. Move forward in executing a plan based on all the knowledge you gathered in Chapter 5. Remember that you need to get to know the blogger, her interests and her audience before reaching out to her. It's important to be transparent. If this is the first time you've ever worked with bloggers, tell her so. She will most likely appreciate your trust in using her blog for testing grounds and offer some of her own unique ideas to maximize your efforts. Visit your ten mom bloggers' site often, watching the comments posted about your product and taking note of other mothers who may potentially join your next effort. A mom who comments about her own positive experience with your brand is a great target for your next blogger outreach.

Remember not all page views and audience sizes are created equal. Make sure you speak to this as you quantify the results of your test program to upper management. Last but not least, ask the mom bloggers who partnered with you for their suggestions and recommendations for future programs. This is the first step to creating an on-going relationship with these women. You can apply the same sense of experimentation to Mom Mavens, vloggers and v-casting. Utilize some existing B-roll of your product and upload it to video sites such as YouTube and Newbaby.com for moms to watch. Allow mothers to post comments and reviews on the video. It's okay for even large global companies to take baby steps in the Mom Market. In fact, it's recommended. Just remember, however, the larger the social reach of your Mom 3.0 initiatives, the greater the benefit in sales, customer acquisition and mom relationships.

For companies with a higher tolerance for more intricately produced marketing programs, there are many ways to approach "*Mom 3.0izing*" your old marketing strategy. For example purposes, let's say that our global marketer represents a food product. Their ultimate marketing campaign makeover would look something like this.

The first thing we will do is create a Mom's Panel. The panel will be comprised of moms who are loyalists to the brand. They will be selected based on essays about their experience with the products. Topics might include "How X product saved me from a parenting disaster" or "How I use X product to create solutions in my family" or even "The first time I ever used X product." Traditional product ads in parenting publications will carry a starburst announcing the call for applications online. Every mom who enters will be mailed a special thank you coupon and/or product sample complete with promo code for tracking purposes.

Once the mom's panel is selected, these women will be used as a sounding board and focus group. Utilize your existing website to link to a micro-site which features the stories and profiles of your Mom's Panel and include interactive features such as blogging and vcasts. Since it's a food product, allow other moms to submit their multiple uses for your product. Simultaneously, launch a mom's e-newsletter that features some of the submissions you receive on the micro-site. We know moms love to share pictures and recipes. Figure out how you can leverage both. Perhaps you allow moms to upload their videos preparing their favorite recipes. Now you are delivering consumer generated content in a very forward minded manner.

Provide a branded widget that allows moms who upload their videos to your site to share with friends and family members. Showcase the Mom's Panels by allowing them to answer questions from readers and have them blog about interesting aspects of your brand.

Meanwhile elsewhere on the web, identify a group of mommy bloggers that fit your market and are willing to do product reviews or run mom promotions. Create a fun package highlighting your product and perhaps the ideas of the moms on your microsite. For instance, if your product is a fast boiling pasta and moms have uploaded videos using your pasta in Mexican, Italian and Oriental dishes, your blogger product kit can carry a theme such as "Around the globe at dinner time with X pasta." In the box you might include the recipes of Susan of Provo, Utah for Mexican Pasta Casserole, Jen of Palo Alto, CA for Traditional Tomato Baked Ziti and Cate of New York, NY for Oriental Pasta Toss. Along with the pasta and recipe cards, include some accessories to create a week of International family dinners. These items can include chop sticks, napkins in Mexican colors, fortune cookies or biscotti for dessert. For her child, you might include dinner conversation cards about the respective countries. Don't forget a few trackable coupons so she can share them with friends. Demonstrate that you are in-tune to the needs of mom bloggers and offer to send each mommy blogger two of your exclusive International Dinner Paks. One she can enjoy with her family and the second she may use to drive traffic to her blog. Leverage her creativity and allow her to create whatever kind of contest or promotion she likes around the second dinner pak. Also offer her the use of your branded widget to share product video or other mom commentary or the special opportunity to interview one of your Mom Panel members.

Thus far you have demonstrated your willingness to help build her brand and given her four ways to interact with you: product review, blog contest, widget and interview. Go the extra step in letting her know you understand her by offering to mail the 2nd basket to her winner for her. Believe it or not, many of these moms do not have the dispensable income to pay for postage so as a sign of best intentions mail the prize basket from your offices to the winning mom. This also allows you to control the presentation of your product.

Utilize television and radio to announce all the mom generated recipes or cooking videos and reinforce your brand messages. For online TV outlets or YouTube, utilize experts from your testing kitchens to create "how-to" content for moms. Speaking of "how-to" content and since you are a master of meal planning solutions, why not create a podcast around the theme of feeding the family? By keeping the subject broad, you can feature topics ranging from nutrition to quick food solutions.

Now that the brand has integrated its traditional media with inactive online media, it's time to enlist the partnership of Mom Mavens offline. Take the content of the International Dinner Packs that you created for your blogger initiative (recipe ideas, coupons and dinner discussion cards) and create a maven mailing. Use social networking sites to put out the word that you are looking for moms with your pre-determined criteria to share product information with friends and peers. Contact some organizations that have local chapters of moms such as Strollerstrides.com or iPlaygroups.com and see if you can partner with their mom leaders. Once you have gathered your database of Mom influencers, you have two great options. The first method is to send them an individual maven mailing with product information, recipe ideas, coupons and dinner discussion cards. This she can enjoy with her family and potentially share the experience with friends. Alternatively, you can enlist her to host an International Dinner party by sending her everything she needs to host a dinner with your product for ten of her friends. Don't forget to follow up with these moms in order to measure your success.

See how easy it is to integrate new social media into old media buying? There is so much to be gained even beyond the valuable relationship you are establishing with mothers. In the scenario I described above your company is gaining access to a built-in focus group, content that can be used in custom publications, television commercials, print ads and online destinations and a database of mothers who are interested in your product. Additionally, you are putting your brand in places where it would not normally be found like in an iTunes directory or as a widget on Grandma Jones's desktop. You are enabling moms to host branded events in their dining room and engaging them to share their own ideas about your product with other moms online. The relationship has so many connection points that it allows Mom 3.0 to customize and personalize the relationship she wants to have with your marketing efforts. Welcome to marketing with Mom 3.0 with old and new mediums.

Start- up companies with limited budgets can also take advantage of all the opportunities of the Mom Market. Expensive print ads can be substituted by forming strategic partnerships with online content sites owned by other mothers. These women are always looking for content so try writing an article that contains mentions of your product. If you created a new line of skid-proof shoes for toddlers, write an article titled, "Avoiding Five Household Dangers Your Walking Toddler Faces." Of course, in the article you will mention slips and falls and the solution you present might be your product. The article should contain a link to your site. In return for placement of your content, offer to donate product for an online sweepstakes or reader contest and, of course, link your site to the posting websites. The cross linking will help both sites optimize search.

You may also want to contact local offline parenting publications regarding your content. Just about every city has their version of a parenting magazine. Particularly smaller publications welcome packaged content because it saves them money for writers and/or syndicating services. The best part is that research shows moms are more likely to purchase a product based on a magazine article than print advertising. Chances are that you don't have expensive television to leverage but have no fear; it's the perfect time to launch your own television network online. An inexpensive Flip camera allows you to create your own BRAND NAME TV. You can upload your online television segments to several other platforms such as YouTube. Since the other recommended initiatives such as Mom Mavens and in home parties are scalable to any budget, integrate them into your plan. Additionally, a strong public relations effort is particularly important when you are launching a brand. It not only allows you to gain credibility among mothers, but also among industry leaders and potential partners. Everyone likes to align themselves with the hottest new product, brand or technology so establishing your place among peers is also vital to your growth.

The key to integrating Mom 3.0 into your 2.0 marketing plan boils down to a few key elements: engaging moms to interact with your brand, providing moms with the tools to share your marketing messaging once she's consumed it and providing her a platform to nurture her own relationship with your company as well as the relationship between you and her peers. Finally, don't forget to provide her the opportunity to share her own ideas with you about your product and/or your marketing campaigns.

The possibilities are endless when it comes to marketing with today's moms. So endless that I could fill a dozen more pages with a flow of ideas, however, all things must come to a close; including this book. It's at this point, however, that I struggle. My previous two books on the Mom Market concluded with a forward look at what's to come in marketing channels and new media. In *Marketing to Moms*, I forecasted a day when moms would have access to an online library of solutions similar to those found in parenting books and today we have Newbaby TV, ParentingTV and Cleverparents TV. In *Trillion Dollar Moms*, the forward thought was a day when brands produced television shows and published their own magazines with partner content. Fast forward to any local Wal-Mart store and moms can pick up a copy of the retailers women's magazine "All You" or view Wal-Mart TV in the aisles. My challenge as I try to wrap up this book is how to be forward thinking in the final chapter of a book that focuses on emerging media and every evolving technology? I think you conclude it by challenging your reader to be adventurous and creative to fully take advantage of all the opportunities that these new marketing channels offer to the bottom line. It's an exciting time in marketing with moms! It's a time like no other in the two decades that I've personally been immersed and focused on mothers. By seizing the moment and riding the momentum of the market, you and your business will find a place of multi-dimensional growth. The relationships you'll cultivate with your customers as they become your marketing partners will provide a rewarding experience for you professionally and most likely, personally. The increased impact of your marketing programs will be felt in sales, retention and brand awareness. Finally, marketing with Mom 3.0 will propel your marketing efforts to a new level of success that will transcend your product, brand or company well beyond the 2.0 world.

Acknowledgements

A DAY DOESN'T GO BY THAT I DON'T RECOGNIZE THAT I'M THE LUCKIEST woman in the world. Personally, I am surrounded by a supportive group of family and friends. Professionally, I have the opportunity to follow my passion and work with the greatest brands around the globe.

Thank you to my clients who make work fun and exciting every single day. I feel privileged to be trusted with your brands, products and services. How many people in life get to work with top brands such as Disney, HP, Precious Moments, Lands' End, HSBC Bank, Primrose Schools, and Cartoon Network among many others? Combined with the chance to know successful entrepreneurs such as Maxine Clark of Build-A-Bear Workshop, Lisa Druxman of Stroller Strides and Gabrielle Blair of Kirtsy.com, the lessons I've learned at each stop are invaluable.

I am also fortunate to be surrounded by what I consider the greatest group of co-workers. Never have I met a more dedicated and committed team when it comes to over delivering on results for clients. Thank you to the women that make up BSM Media: Laura Motsett, Valika Shivcharran, Amy Shiman, Natalie Zupo, Heidi Fliess, Holly Edger, Carlene Wegmann Todd, Lisa Shaw and Jennifer Arnold. I cannot say enough about the fun and hard work each of these women bring to the job every single day. Thank you so much for bringing my marketing ideas to life and producing the results necessary to grow them. And of course, Judene Somers who is always there to support BSM Media. In addition to my real job, I work every day with a radio crew that starts the day off with laughs and thought-provoking discussion. I am so fortunate to be part of the Doug Stephan's "Good Day" show and I thank Doug immensely for the opportunity to be a part of the greatest morning talk show on national radio today! Thanks also to Rich McFadden, Jennifer Horn and Roberta Facinelli who put up with my conservative on-air comments. Speaking of radio, thank you to the guy who makes Mom Talk a reality every week—my producer Mike McGann. You are the best! Finally, thanks to my two business partners, Rachael Bender and Bob Sullivan. Rachael and I created BlueSuitMom.com almost a decade ago long before podcasts, blogs and video content. She's the smartest person I've ever met and her continued support has been something I could not live without. Bob and I created Newbaby.com. We met by accident, but the outcome has revolutionized the delivery of video content to new moms. Thank you Bob for the ride!

I am grateful for the friendships I've been able to establish because of my extensive travels. People that I wish I had more time to share a Cosmo, long walk or extended dinner with. Thank you to this special group of friends: Nancy Cleary, Susan Meek, Mike Hyland, Karen Cage, Victoria Naffier, Gwen Mahoney, Leanne Jakubowski, Carter Auburn, Katherine Durham, Catherine O'Hare, Bret Moore, Julie Fuoti, Allison Zane, Mia Cronan, Kit Bennett, Cindy Robinson, Jeffrey Vinghoff, Karen Deerwester, Jo Kirchner and Paul Thaxton. The list is too long to include so I will thank all of my Mommy Blogger friends who support my ideas and keep me informed on what's new in the Blogosphere. To my fellow founders of The Marketing to Moms Coalition: Bridget Brennan, Amy Colton, Teri Lucie Thompson and Michal Clements. Thank you for bringing to life a dream for an industry organization to fuel the fires of Marketing with Moms. Thank you also to my girl friends, Audrey Ring, Jennifer Calhoun, Alice White and Brenda Kouwenhoven. You're always there to give my kids a ride home, keep me up to date on school activities or listen to my troubles. I appreciate your support so very much.

Thank you to Duncan Wardle, Maxine Clark and Byron Norfleet for endorsing my book. A special word of appreciation to Michael Mendenhall for agreeing to write the Foreword. You are one of the most creative, intelligent and super-human marketers I've ever met and I am honored to have your insights grace the pages of *Mom 3.0*. Thank you so very much.

To my family who still can't figure out what I do for a living, thank you for not calling my cell phone at 5:00 a.m. when I'm in California or asking me when I am going to stop working so much. Thanks to my brothers, Matthew and Michael John for keeping me happy with fishing trips and lots of fish. And to my other brothers Mike and Bryan, thanks for keeping me laughing at family dinners. To my sister Debbi, I appreciate your friendship more than you will ever know. Thank you to my parents Jackie Alligood, Bill and Susan Telli and the late Dr. H. Michael Alligood. I wouldn't be who I am without the sacrifices you made for me. All this leads to the most supportive and tolerant group of family members in my life, my husband and children. You give up so much in order for me to follow my passion. I am more than proud of you and hope that you will learn from the example I set that your career can be fun, rewarding and fulfilling. I am indeed the luckiest mom in the world to have you in my life.

INDEX

Printed in the United States
125657LV00002B/37/P

9 781932 279900